exploring **mathematics** with scientific notebook®

Springer
Singapore
Berlin
Heidelberg
New York
Barcelona
Budapest
Hong Kong
London
Milan
Paris
Santa Clara
Tokyo

exploring mathematics with scientific notebook®

Wei-Chi YANG and Jonathan LEWIN

Dr. Wei-Chi Yang
Department of Mathematics and Statistics
Radford University
Radford, VA 24142
USA

Prof. Jonathan Lewin
Department of Mathematics
Kennesaw State University
1000 Chastain Road
Kennesaw, GA 30144-5591
USA

Library of Congress Cataloging-in-Publication Data

Yang, Wei-Chi, 1959-
Exploring Mathematics with Scientific Notebook/ Wei-Chi Yang, Jonathan Lewin.
p. cm.
Includes index.
ISBN 9813083883
1. Scientific Notebook. 2. Mathematics--Data processing.
I. Lewin, Jonathan.
QA76.95.Y36 1998
510'.78' 55369—dc21 98-10227
CIP

ISBN 981-3083-88-3

Printed in Singapore

Typesetting: Camera-ready by Authors
SPIN 10671201 5 4 3 2 1 0

Preface

This book contains a number of separate modules that are designed to serve as supplements to the standard bill of fare in the study of algebra, calculus, linear algebra, analysis and other topics in an undergraduate mathematics curriculum. A central thrust of these modules is that they exploit the special features of the software products *Scientific Notebook* and *Scientific WorkPlace 3.0* and, in doing so, they suggest a variety of innovative approaches to the subject matter that can be adopted with the help of these important software tools. The process of reading the material of this book is the same in *Scientific WorkPlace* and *Scientific Notebook* and, from now on, when we refer in this book to *Scientific Notebook*, the reference should be understood as applying equally to *Scientific WorkPlace* and *Scientific Notebook*.

This book is supplied both as a traditionally printed and bound textbook and as an on-screen version that you will find in the compact disk that is included with the book. It is the on-screen version of the book that the reader will find most useful. This on-screen version will be updated from time to time and will include hypertext links to documents that have been placed in the World Wide Web or in FTP sites. One advantage of such links is that the target documents can be updated at any time.

Each module of this work is intended to be read on the computer screen and it takes advantage of the instant access that *Scientific Notebook* provides to the computing engine Maple (a product of Waterloo Maple Software Inc.) Since the link to Maple provided by *Scientific Notebook* enables the student to work with Maple without having to know any of its special syntax, the modules in this book can be read in a truly interactive environment.. The modules encourage the student to experiment with the material and to explore a variety of possibilities and "what if" situations that may come to mind. Some of the modules contain external program calls to the Windows media player. With a single click of the mouse, a reader can play animations of two or three dimensional graphs and gain greater insight than can be provided by graphs that are drawn one at a time. Occasionally, a module may contain an external program call to a direct session of Maple or Mathematica (a product of Wolfram Research Inc.) for the benefit of readers who have their own stand-alone versions of those products.

This book is not designed to be read from cover to cover in any one course. Instructors are encouraged to scan the book for the chapters (modules) that are relevant to the course for which they wish to use this work and then to delete the chapters that they will not be using before they supply the material to their students. Most instructors will choose to begin with the first chapter that is designed as a basic boot camp in the use of *Scientific WorkPlace* to perform mathematical operations. The following list of chapter headings provides a quick look at the subject matter of this book. To jump to the beginning of any chapter, point at the chapter name in this list, hold down the control key and click the left mouse button.

Reading the Hard Copy Version of this Book

As indicated in the preface, this book is meant to be read on-screen with the aid of *Scientific Notebook* or *Scientific WorkPlace 3.0*. Although the hard copy version of this book is an indispensable part of it, there are some items that simply can't be read in the hard copy. Among these are the following:

1. Mathematical operations that are performed by Maple as the book is being read.
2. Animations of some of the pictures that are achieved in the on-screen version by clicking on an external program call to the media layer.
3. Certain displayed expressions that will not fit into a single line. These have been truncated in the hard copy and can be seen in their entirety only in the on-screen version.

Acknowledgments

The remarkable cover of this book was made from a photograph of an original pen and ink drawing of Ariel Lewin. We would like to express my special thanks to her for permission to use one of her creations. Ariel can be reached by e-mail at lewins@mindspring.com by anyone who would like further information about her work.

The authors would like to express their sincere appreciation to their families who provided moral and emotional support and shouldered a wide variety of chores to make it possible for us to work on this text.

We would like to express our deep appreciation to Roger Hunter, the designer of the software products *Scientific WorkPlace* and *Scientific Notebook* without which this book could not have been written. Roger and his dedicated team at TCI Software Research have provided the mathematical community with a priceless asset.

We would like to express our appreciation to the Brooks/Cole Publishing Company and to International Thomson Publishing, the parent companies of TCI Software Research for giving us permission to include their software on the CD that is provided with this book. In particular, we would like to express our appreciation to George Pearson of TCI Software Research for creating the master CD.

Last, but not least, we would like to express our appreciation to the folks at Springer-Verlag for the excellent job they have done in the production of this book.

Wei-Chi Yang
Radford University

Jonathan Lewin
Kennesaw State University

Contents

Chapter 1
Reading this Book On-Screen

This book is supplied both as a traditionally printed and bound textbook and as an on-screen version that you will find in the compact disk that is included with the book. Although some features of this book are designed to be read in the traditional way, you will be able to gain much more value from this book if you are can read it on-screen as well.

1.1 What Do I Need to Read this Book On-Screen?

The on-screen version of this book is designed to be read with the software product *Scientific Notebook* that is made by TCI Software Research which is a division of the Brooks/Cole Publishing Company. Alternatively, you can use the flagship product, *Scientific WorkPlace*, of TCI Software Research. In order to run *Scientific Notebook* in your computer you need to be running Windows 95 (or Windows NT 4.0).

For the purpose of reading this book, *Scientific Notebook* and *Scientific WorkPlace* can be thought of as acting in exactly the same way and a reference to *Scientific Notebook* in this text should be taken as referring either to *Scientific WorkPlace* or *Scientific Notebook*. You can install *Scientific Notebook* from the compact disk that is provided with this book. Alternatively, you can download it from the internet. Please visit

http://www.scinotebook.com

where you will also be provided with information about *Scientific Notebook* and *Scientific WorkPlace* and the option to download *Scientific Notebook*. When installing *Scientific Notebook*, you will have two options:

1. You may install a reader version of *Scientific Notebook* that will allow you to read this book on-screen and will work permanently in your computer. However, this reader version will not allow you to save any changes that you make in documents and will not provide you access to the computing features of Maple.
2. You may install the full version of *Scientific Notebook* and use it on a trial basis for 30 days. After this time, you can purchase *Scientific Notebook* for just $60 to make its installation permanent in your computer.

1.2 What is *Scientific Notebook*?

Scientific Notebook is a combination word processor and computing product designed for use under Windows 95 or Windows NT 4.0. It combines the features of a powerful and friendly scientific word processor with the computing features of the computing engine Maple[1] which is included with *Scientific Notebook*. This link to Maple makes it possible for a *Scientific Notebook* user to work directly with mathematical expressions that have been written into a document; expressions which have exactly the same form as those that one would write with pencil and paper. There are no commands to learn, nor any syntax to remember. Furthermore, the mathematical expressions upon which the operations are being performed are integrated seamlessly into the document. Since *Scientific Notebook* does not require the use of an "equation editor", there is no need to place mathematical expressions in a separate window.

This book is not meant to provide a complete training in the use of *Scientific Notebook* to do mathematics. You can find extensive coverage of this topic in the following books:

- *Doing Mathematics with Scientific WorkPlace* by Darel Hardy and Carol Walker (provided with *Scientific WorkPlace*).
- *Doing Calculus with Scientific Notebook* by Darel Hardy and Carol Walker, published by Brooks/Cole Publishing Company.
- *Doing Linear Algebra with Scientific Workplace* by Manfred Szabo to appear.
- *Precalculus with Scientific Notebook* by Jonathan Lewin, published by Kendall/Hunt Publishing Company, 1997.

The purpose of this chapter is to give you some quick training in the use of those features of *Scientific Notebook* that you will need to read this book and a brief overview of some of the computing features.

1.3 Getting Started

1.3.1 Installing *Scientific Notebook*

Place your CD in your CD-ROM drive and close the door. You will see the installation screen.

Click on Run to begin the installation.

1 Maple is a product of Waterloo Maple Software.

1.3.2 Installing the Book in Your Computer

Place your CD in your CD-ROM drive and close the drawer and, when you see the installation screen, click on Exit. Now use your Windows Explore to examine your CD. You will see a folder named YL-Book there. Copy this folder to your hard drive. Now open *Scientific Notebook* and click on the option File at the top left of your screen.

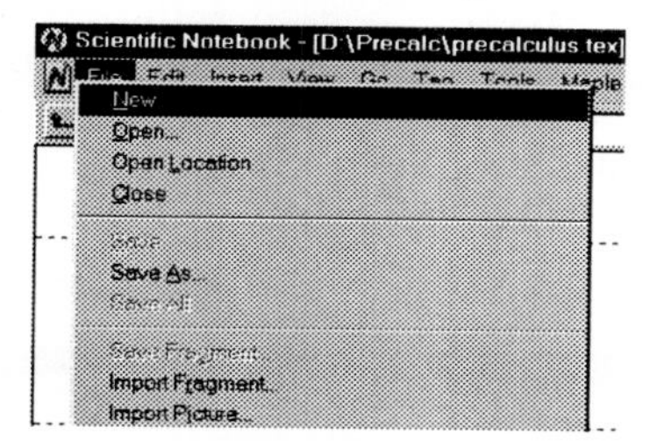

When the file menu opens, click on Open. You will see the file open dialog box. Browse to your new folder, select the file yl-book.tex and then click on Open.

1.4 Updating Your Document

From time to time, the on-screen version of your book will be updated. You can find updates at the location

http://science.kennesaw.edu/~jlewin/ylbook

The file names will be yl-bookxxx.zip where xxx stands for the current version number of the manuscript. These files will be compressed with PKZip. To view the files in the yl-book folder, follow the instructions in this part of the on-screen version of the book.

1.5 Navigating in this Document

In order to read this book efficiently on the computer screen you have to be able to scroll through it quickly and painlessly. As in most Windows products, you can use arrow keys to scroll through your documents line by line, the Page Up and Page Down keys to move up and down one screen at a time and the vertical scroll bar at the right of your screen to make larger jumps. In addition to these traditional ways of moving around in the document, *Scientific Notebook* also provides some special navigation features. The simplest way to access these special features is to use the Navigate and Link toolbars.

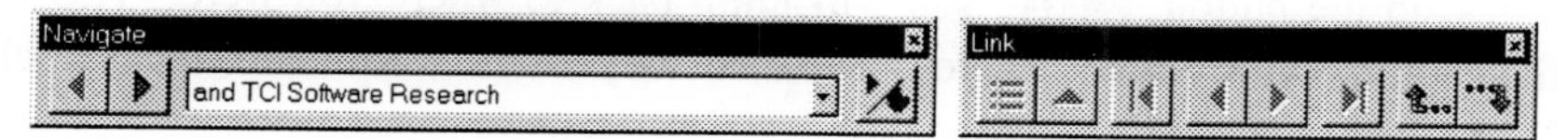

If these toolbars do not appear in your screen, click on the View menu at the top of your screen. When it opens you will see

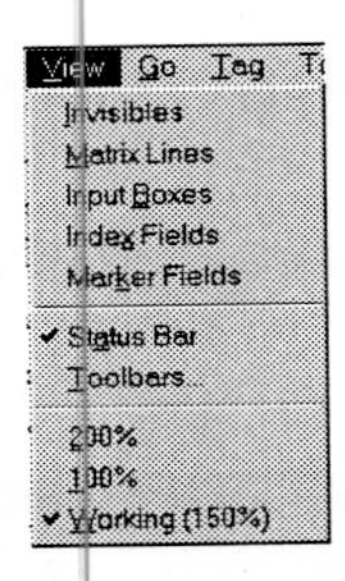

Click on Toolbars and you will see the toolbar menu

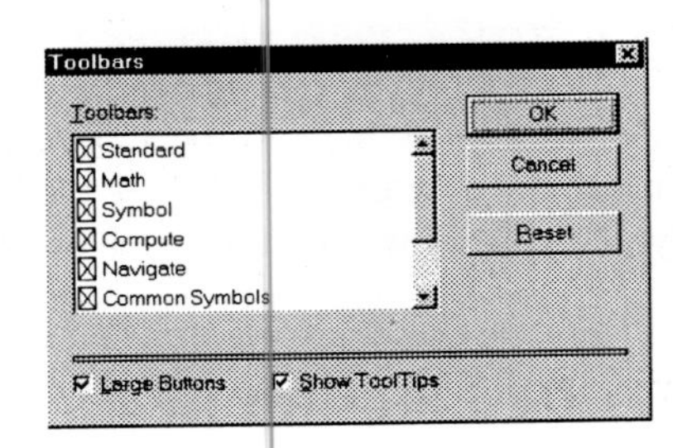

Make sure that the toolbars you want to see are checked. Note that in the above figure, the option Large Buttons has also been checked. You will want large buttons only if your Windows session is running with a high screen resolution.

Click on the down arrow in your Navigate toolbar. You will see a table of contents for the document you are reading.

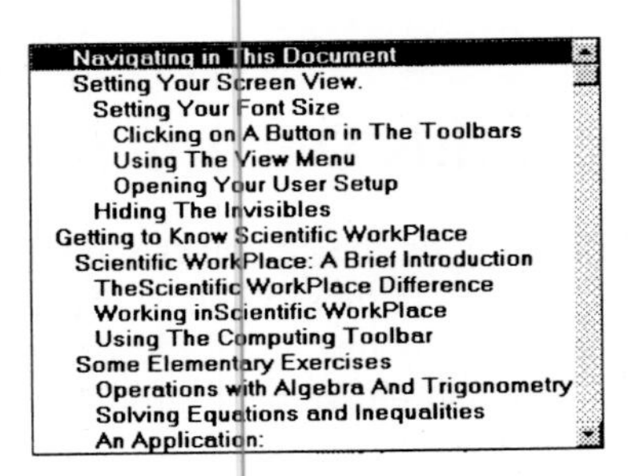

You can scroll through this table of contents and click on any item in order to jump to it. If you click on the button you will jump back to where you were. This button is called the History back button. There is also a History forward button. You can click on the buttons and to move forward or backward from heading to heading. Finally, you can click on the button to open up the Go to Marker menu

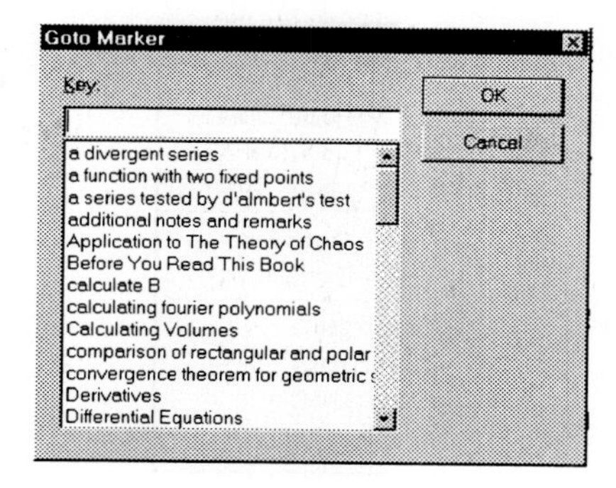

which shows all of the markers in your document in alphabetic order. This list of markers plays the role of an on-screen index and allows you to jump to any item with a single mouse click.

1.6 Setting Your Screen View

Your enjoyment of the on-screen version of this book will be increased if you optimize your screen view. This section contains some suggestions that you may use for this purpose.

1.6.1 Setting Your Font Size

The ideal size for the screen fonts in your particular installation of *Scientific Notebook* depends upon the properties of your computer and the way it is configured. For example, if your computer is configured for a fairly high screen resolution then you will need to choose fairly large screen fonts to view your documents comfortably. Please note that adjusting your screen fonts is done only for your comfort in reading documents on the computer screen. In no way will a change in the size of your screen fonts affect the way in which your printed documents appear.

There are two main ways in which you can adjust the size of your screen fonts.

- If you have chosen to show the Standard toolbar then you will see the button 150% The size of your present screen font is shown there. By clicking on the little down arrow on the right of this button you can bring up a menu of other screen font sizes. You can change to any of these by pointing at it and clicking the mouse and you can also type in your own size selection.
- At the top of your computer screen you will see several menus, one of which is labelled View. Click on this menu and it opens showing

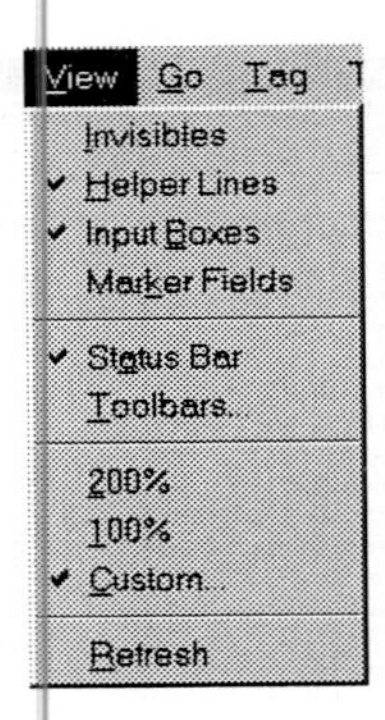

and you can choose quickly between any of the three screen font sizes shown in this menu by pointing and clicking. If you choose Custom you will be presented with the opportunity of typing in your size selection.

1.6.2 Hiding the Invisibles

Invisibles are formatting characters and other objects that we sometimes need to see on the computer screen even though they would not appear in the printed form of the document. In general, it is a good idea to show at least some kinds of invisibles while you are writing a document but you should hide them when you are reading a document that has already been created. To make sure that all of the invisibles are hidden while you are reading this book, open the View menu again and make sure that each of the items Invisibles, Matrix Lines, Input Boxes, Index Fields, and Marker Fields is unchecked. If any of these are checked, you can uncheck the item by pointing at it and clicking the mouse. Even if you decide that you would like some of these invisibles to show, you should probably hide Marker Fields since these marker fields can be quite ugly.

1.6.3 Working with Read-Only Documents

If you are working with a read-only form of this book you may not be able to make changes to it. In this case you have several options:

1. Open a new document for your notes, comments and experiments. Remember that you can have several documents open at any one time and switch between them by clicking on Window.
2. Open a new document with a book style (such as the Standard LaTeX Book Style) and then click on File and Copy Contents in order to copy the contents of this book into your own document. Your new document will no longer be read-only.
3. Use your operating system's file management system to remove the read-only attribute from your copy of this book. This option will not be open to you if the computer is in a computer lab. If you do decide to make changes in the original form of this book, be sure you keep a backup copy in case you damage the document.

1.7 The Graphs in this Book

Graphs drawn in a *Scientific Notebook* document are living things. Every time you open a document which contains a graph a graph you have drawn, Maple will redraw the graph and will create a new snapshot of it. This feature of *Scientific Notebook* allows us to revise graphs at any time. We can change their colors, their domains, their orientations and many other properties. We can even add new plots to graphs that have already been drawn.

On the other hand, when Maple has to redraw graphs very frequently, the extra load this drawing places on your computer may be very noticeable, especially if you have a slower CPU. We have therefore chosen not to leave the graphs drawn in this book as actual graphs. Instead, we have provided pictures that were made with the snapshots that Maple created when the graphs were drawn. Even these pictures will sometimes cause your computer to pause momentarily as you scroll to them but they are faster then the graphs themselves would have been.

It is strongly recommended that, as you read this book, you should redraw the graphs yourself following the instructions that will be provided for you. In this way you will be able to experiment with the graphs, add plots to them and change their properties. When you look at 3D graphs you will certainly want to rotate the graphs in order to view them from the most pleasing angle.

1.8 Reading and Writing in *Scientific Notebook*

The text editing component of *Scientific Notebook* is a word processing system that has the ability to distinguish between **text** and **mathematics**. At any time, *Scientific Notebook* is either in **text mode** or it is in **mathematics mode**. You can see which mode is active at any given time by looking at your **mathematics toolbar**

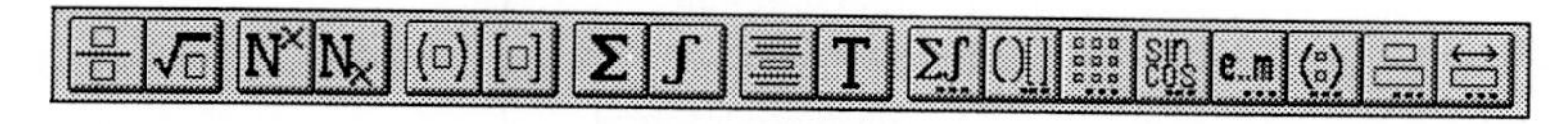

or your **standard toolbar**

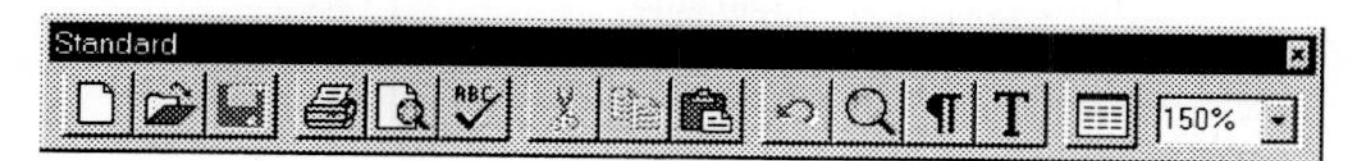

depending upon your version of *Scientific Notebook* or *Scientific Notebook*. If the button T is showing there then *Scientific Notebook* is in text mode and any symbols you type will be treated as text. Such symbols will usually be black but, depending on their position in the document and the typeface being used, they may also appear in other colors. If the mathematics toolbar shows the symbol M then *Scientific Notebook* is in mathematics mode and any symbols you type will be typeset according to mathematical conventions and also recognized as mathematics for the purposes of mathematical operations. There are many ways to change the mode of *Scientific Notebook* from text to mathematics and back again. One way is to point the mouse at the symbol T and click it to change to mathematics mode and to point at the symbol M and click to change to text mode.

Mathematical operations are performed in a *Scientific Notebook* document by the computing engine Maple which is supplied with *Scientific Notebook*. If you click on the Maple item at the top of your screen you will see the pull-down menu:

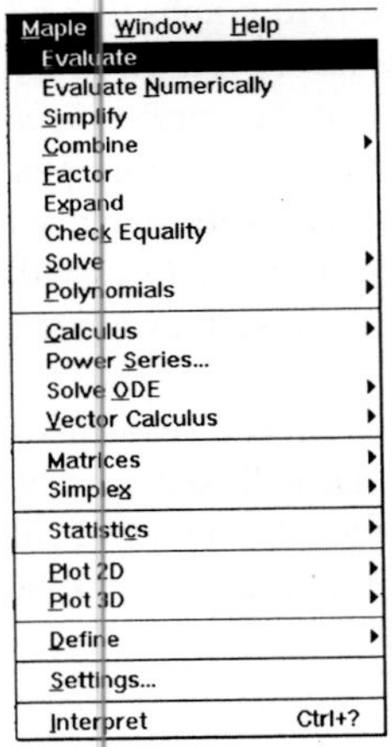

Maple Window Help
Evaluate
Evaluate Numerically
Simplify
Combine
Factor
Expand
Check Equality
Solve
Polynomials
Calculus
Power Series...
Solve ODE
Vector Calculus
Matrices
Simplex
Statistics
Plot 2D
Plot 3D
Define
Settings...
Interpret Ctrl+?

which gives you a brief listing of the operations you can perform with Maple. As you can see, many of the items have submenus that can be opened by clicking on the arrow to the right. For example, clicking on the item Calculus brings up the submenu:

Integrate by Parts...
Change Variable...
Partial Fractions
Approximate Integral...
Plot Approx. Integral
Find Extrema
Iterate...
Implicit Differentiation...

Every one of the Maple operations can be performed by the simple operation of pointing at one of these menu items and clicking. Thus, for example, if you place your cursor in the expression $\frac{5}{12}+\frac{2}{15}$ and then click on the item Evaluate in the Maple menu you will see

$$\frac{5}{12}+\frac{2}{15}=\frac{11}{20}.$$

Thus, as we have said, you can perform mathematical operations in *Scientific Notebook* without having to learn any programming language. Every operation is performed upon a mathematical expression that looks just like one you would have written with pencil and paper.

1.9 Using the Computing Toolbar

Depending upon how you have your copy of *Scientific Notebook* set up, you may or may not have chosen to view the **Computing toolbar.**

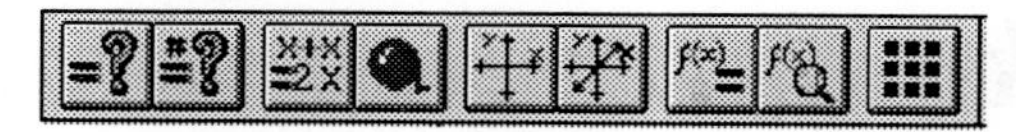

If you want to view this toolbar, click on View and then on Toolbars. Once the computing toolbar is in view you can use it to obtain shortcuts to many of the common mathematical operations. Point the mouse at the buttons to allow the yellow tool tips to show. You will see, for example, that the button gives Evaluate, the button gives Evaluate Numerically, the button gives Simplify and the button gives Expand.

There is yet another way of selecting Evaluate. Point at any expression that you want to evaluate, hold down the Control key and then press e.

Chapter 2
Getting to Know *Scientific Notebook*

2.1 Computing Exercises with *Scientific Notebook*

In this section you will see a variety of exercises that will help you learn how to take advantage of the computing power of *Scientific Notebook*. Work through as many of these exercises as possible but ignore any that use mathematical notation that you have not seen before.

2.1.1 Operations with Algebra and Trigonometry

1. Point at the expression 2^{38}, open the Maple menu and click on the item evaluate. You will obtain $2^{38} = 274877906944$. Try this exercise again using the items simplify and expand.
2. Highlight the expression 2^{38} and, while holding down the control key, click on Evaluate. The expression 2^{38} will be replaced by 274877906944.
3. Point at the expression $\left(\frac{2}{3}\right)^{38}$ and click on evaluate. You will obtain

$$\left(\frac{2}{3}\right)^{38} = \frac{27\,48779\,06944}{1350\,85171\,76729\,92089}.$$

However, if you click on Evaluate Numerically you will obtain

$$\left(\frac{2}{3}\right)^{38} = 2.034848854 \times 10^{-7}.$$

4. Use Evaluate Numerically to show that $\pi = 3.141592654$.
5. Open the Maple menu, click on Settings to bring up the menu

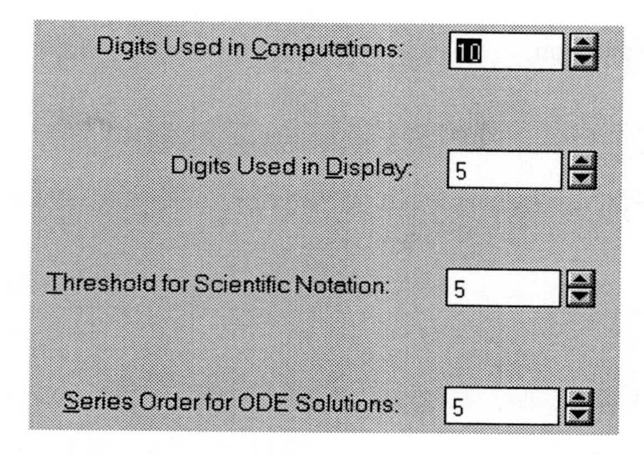

and adjust the number of digits to show that $\pi = 3.14159265358979323 85$.

6. Point at the expression $\cos 30°$ and click on Evaluate to obtain $\cos 30° = \frac{1}{2}\sqrt{3}$. Do this again with Evaluate Numerically to obtain $\cos 30° = .86603$.

7. Point at the expression $\cos 30$ and click on Evaluate Numerically to obtain $\cos 30 = .15425$.

8. Point at the expression

$$\frac{2}{x-1} - \frac{3}{x+1}$$

and click on Simplify to obtain

$$\frac{2}{x-1} - \frac{3}{x+1} = -\frac{x-5}{(x-1)(x+1)}.$$

9. Point at the expression

$$\operatorname{lcm}(yx + 3x - 5y - 15, xz - 53x - 5z + 265)$$

and click on Evaluate to obtain

$$yxz - 53yx - 5yz + 265y + 3xz - 159x - 15z + 795.$$

Then click on Factor and obtain

$$(z - 53)(y + 3)(-5 + x).$$

10. Point at the expression

$$\left(1 - x + 2x^2\right)^{12}$$

and then click on Expand. Then point at the answer and click on Factor to obtain $\left(1 - x + 2x^2\right)^{12}$ again.

11. Point at the expression $a^3 + b^3 + c^3 - 3abc$ and click on Factor to show that

$$a^3 + b^3 + c^3 - 3abc = (a + c + b)\left(a^2 - ac - ab - bc + c^2 + b^2\right).$$

12. Point at the expression

$$\frac{x^8}{x^2(x-2)^3(x^2+x+1)}$$

and click on Polynomials and then on Partial Fractions to obtain

$$\frac{x^8}{x^2(x-2)^3(x^2+x+1)} = x+5+\frac{64}{7(x-2)^3}+\frac{1024}{49(x-2)^2}+ \frac{6192}{343(x-2)}-\frac{1}{343}\frac{19+18x}{x^2+x+1}$$

13. Point at the expression $\log_{10} 100$ and click on Evaluate to obtain $\log_{10} 100 = 2$.
14. Point at the expression $\log 100$ and click on Evaluate Numerically to obtain $\log 100 = 4.6052$. The point of this exercise is to emphasize that the symbol log standards for the natural logarithm in *Scientific Notebook*. However, if you prefer to make log stand for $\log_{10}$ you can request this by opening the Maple menu, clicking on Settings and then clicking on Definition Options. This brings up a dialog box that allows you to choose a meaning for the symbol log. The pertinent part of this dialog box appears as

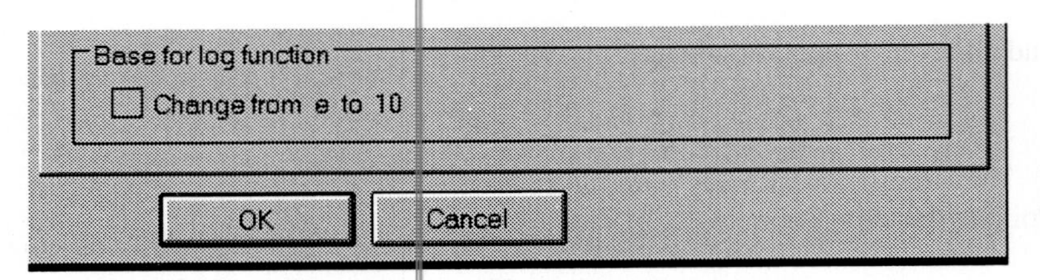

15. Point at the expression $\sin 6\theta$ and click on Expand to obtain

$$\sin 6\theta = 32\sin\theta\cos^5\theta - 32\sin\theta\cos^3\theta + 6\sin\theta\cos\theta.$$

16. Point at the expression $\sin(2a+3b)$ and click on Expand to obtain

$$\sin(2a+3b) = 8\sin a\cos a\cos^3 b - 6\sin a\cos a\cos b + 8\cos^2 a\sin b\cos^2 b - 2\cos^2 a\sin b - 4\sin b\cos^2 b + \sin b$$

17. Point at the expression

$$\frac{\sin 3\theta + \sin 7\theta}{\cos 3\theta + \cos 7\theta} - \tan 5\theta$$

and click on Simplify to obtain

$$\frac{\sin 3\theta + \sin 7\theta}{\cos 3\theta + \cos 7\theta} - \tan 5\theta = 0.$$

18. Point at the expression

$$\arcsin\left(\frac{\sqrt{5}-1}{4}\right)$$

and click on **Evaluate** to obtain

$$\arcsin\left(\frac{\sqrt{5}-1}{4}\right) = \frac{1}{10}\pi.$$

19. Point at the expression

$$\begin{bmatrix} -37 & -35 & 97 & 50 & 79 \\ 56 & 49 & 63 & 57 & -59 \\ 45 & -8 & -93 & 92 & 43 \\ -62 & 77 & 66 & 54 & -5 \\ 99 & -61 & -50 & -12 & -18 \end{bmatrix}^{-1}$$

and click on **Evaluate** to obtain

$$\begin{bmatrix} -37 & -35 & 97 & 50 & 79 \\ 56 & 49 & 63 & 57 & -59 \\ 45 & -8 & -93 & 92 & 43 \\ -62 & 77 & 66 & 54 & -5 \\ 99 & -61 & -50 & -12 & -18 \end{bmatrix}^{-1}$$

$$= \begin{bmatrix} \frac{3646045}{830896983} & \frac{61032898}{830896983} & \frac{17367281}{830896983} & -\frac{121564666}{830896983} & -\frac{108793723}{830896983} \\ \frac{282889}{830896983} & \frac{75453310}{830896983} & \frac{23138783}{830896983} & -\frac{152326234}{830896983} & -\frac{148488790}{830896983} \\ \frac{4369858}{830896983} & \frac{12590134}{830896983} & -\frac{609127}{830896983} & -\frac{20278231}{830896983} & -\frac{17911135}{830896983} \\ -\frac{860894}{830896983} & -\frac{45903257}{830896983} & -\frac{9954523}{830896983} & \frac{102576770}{830896983} & \frac{94408511}{830896983} \\ \frac{836667}{92321887} & \frac{8400848}{92321887} & \frac{2825961}{92321887} & -\frac{18271675}{92321887} & -\frac{17166696}{92321887} \end{bmatrix}$$

now try **Evaluate Numerically** to obtain[2]

$$\begin{bmatrix} 4.3881\times10^{-3} & 7.3454\times10^{-2} & 2.0902\times10^{-2} & -.14631 & -.13094 \\ 3.4046\times10^{-4} & 9.0809\times10^{-2} & 2.7848\times10^{-2} & -.18333 & -.17871 \\ 5.2592\times10^{-3} & 1.5152\times10^{-2} & -7.331\times10^{-4} & -2.4405\times10^{-2} & -2.1556\times10^{-2} \\ -1.0361\times10^{-3} & -5.5245\times10^{-2} & -.01198 & .12345 & .11362 \\ 9.0625\times10^{-3} & 9.0995\times10^{-2} & .03061 & -.19791 & -.18594 \end{bmatrix}$$

[2] The following expression will be fully visible only in the on-screen version of this text.

2.1.2 Solving Equations and Inequalities

1. Point at the equation $x^2 - x - 6 = 0$ and click on Solve and then Exact to obtain: $\{x = 3\}, \{x = -2\}$.
2. Point at the expression $x^2 - x - 6$ and click on Solve and then Exact to obtain: $\{x = 3\}, \{x = -2\}$.
3. Point at the equation $2x^3 - 7x^2 + 5x + 2 = 0$ and click on Solve and then Exact to obtain the solution

$$\{x = 2\}, \left\{x = \frac{3}{4} + \frac{1}{4}\sqrt{17}\right\}, \left\{x = \frac{3}{4} - \frac{1}{4}\sqrt{17}\right\}.$$

4. Point at the equation $x^3 + x + 3 = 0$ and click on Solve and then Numeric to obtain the solution $\{x = -1.2134\}$. What happens if you click on Solve and then Exact?
5. Point to the equation $3^x = 8$ and click on Solve and then Exact to obtain

$$\left\{x = \frac{\ln 8}{\ln 3}\right\}.$$

6. Point at the system of equations

$$\begin{aligned} x - 2y &= 3 \\ 3x - 4y &= 7 \end{aligned}$$

   and click on Solve and then Exact to obtain $\{y = -1, x = 1\}$.
7. Point at the system of equations

$$\begin{aligned} x - 2y &= 3 \\ 2x - 4y &= 6 \end{aligned}$$

   and click on Solve and then Exact to obtain$\{x = 2y + 3, y = y\}$.
8. Point at the system of equations

$$\begin{aligned} x - 2y &= 3 \\ 2x - 4y &= 7 \end{aligned}$$

   and click on Solve and then Exact and observe that nothing happens. Of course, this system of equations has no solution.
9. Point at the system of equations

$$\begin{aligned} 2u + 3v - 7x + y + 3z &= 2 \\ u - 4v + 2x + 3y - z &= 1 \\ -3u + v - x + 2y + 4z &= -3 \end{aligned}$$

   and click on Solve and then Exact and you will see the dialog box

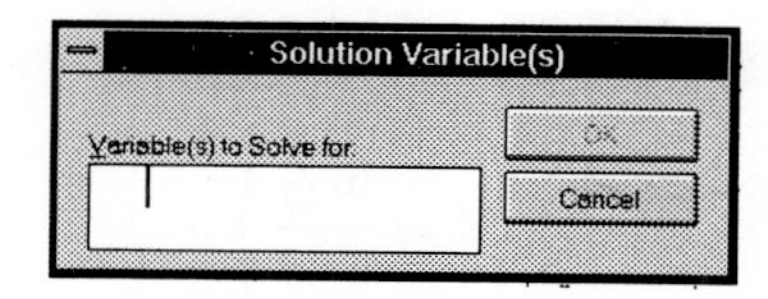

In this box fill x, y, z and then click on OK. You will obtain:

$$\left\{y = -\frac{3}{7}u + \frac{13}{14}v + \frac{3}{7}, z = \frac{8}{7}u - \frac{9}{14}v - \frac{8}{7}, x = \frac{5}{7}u + \frac{2}{7}v - \frac{5}{7}\right\}$$

10. Point at the equation

$$\frac{1}{x} + \frac{1}{y} + \frac{1}{z} = 1$$

and click on Solve and then Exact. Then choose to solve for x to obtain

$$\left\{x = -y\frac{z}{z + y - yz}\right\}.$$

11. Point at the equation $x^3 + 2x^2 - 12x + 8 = 0$, and click on Solve and then click on Exact to obtain: Solution is $\{x = 2\}, \{x = -2 + 2\sqrt{2}\}, \{x = -2 - 2\sqrt{2}\}$.
12. Point at the equation $8x^3 - 6x - 1 = 0$ and click on Solve and then Numeric to obtain $\{x = -.76604\}, \{x = -.17365\}, \{x = .93969\}$. Try this again with Solve and then Exact to see what happens.
13. Point at the equation $x^2 - x\cos x + 2x\log_3 x = 5$ and click on Solve and then Numeric to obtain $\{x = 1.7308\}$
14. Point at the inequality

$$\frac{7 - 2x}{x - 2} > 20,$$

and click on Solve and then Exact to obtain

$$\left\{x < \frac{47}{22}, \ 2 < x\right\}.$$

The comma in this form of the solution means *and.*

15. Point at the inequality

$$\frac{7 - 2x}{x - 2} < 20,$$

and click on Solve and then Exact to obtain

$$\{x < 2\}, \ \left\{\frac{47}{22} < x\right\}.$$

In this solution the comma means *or.*

16. Solve the equation $P = Qe^{kt}$ for k obtaining

$$\left\{k = \frac{\ln \frac{P}{Q}}{t}\right\}.$$

17. Solve the equation $\log_5(4x^2 - 3y) = 5^{\frac{5}{\ln 5}}$ for x obtaining

$$\left\{x = -\frac{1}{2}\sqrt{\left(e^{5\frac{5}{\ln 5}\ln 5} + 3y\right)}\right\}, \left\{x = \frac{1}{2}\sqrt{\left(e^{5\frac{5}{\ln 5}\ln 5} + 3y\right)}\right\}.$$

18. Point at the equation $\sin 4x = \frac{1}{3}$ and click on Solve and then Exact to obtain the solution $\{x = \frac{1}{4}\arcsin\frac{1}{3}\}$. Observe that *Scientific Notebook* gives just one solution of this equation. However, by selecting Solve and then Numeric you can obtain a solution in a specified interval. Point at the equation

$$\begin{aligned} \sin 4x &= \frac{1}{2} \\ x &\in [1, 2] \end{aligned}$$

and click on Solve and then Numeric to obtain the solution $\{x = 1.7017\}$.

19. Point at the equation $x = 10\sin x$ and click on Solve and then Numeric to obtain the solution $\{x = 2.852342\}$.

2.1.3 A Geometric Problem

Two planks, one of length 20 feet and the other of length 30 feet, are crossed in such a way that the plane that contains them is perpendicular to the ground and such that their point of intersection is 8 feet from the ground. Find the distance between the bases of the two planks and the height above the ground of the top of each plank.

2.1.4 Calculus Operations

1. Point at the expression

$$\frac{d}{dx}\frac{x^3\cos\left(3x^2 + x + 1\right)}{(x+1)\ln x}$$

and click on Evaluate to obtain

$$3x^2\frac{\cos\left(3x^2+x+1\right)}{(x+1)\ln x} - x^3\left(\sin\left(3x^2+x+1\right)\right)\frac{6x+1}{(x+1)\ln x}$$

$$-x^3\frac{\cos\left(3x^2+x+1\right)}{(x+1)^2\ln x} - x^2\frac{\cos\left(3x^2+x+1\right)}{(x+1)\ln^2 x}.$$

2. Point at the expression $\int \sqrt{1-x^2}dx$ and click on Evaluate to obtain

$$\int \sqrt{1-x^2}dx = \frac{1}{2}x\sqrt{(1-x^2)} + \frac{1}{2}\arcsin x.$$

3. Point at the expression $\int_0^1 \sqrt{1-x^2}dx$ and click on Evaluate to obtain

$$\int_0^1 \sqrt{1-x^2}dx = \frac{1}{4}\pi.$$

4. Point at the expression $\int_0^1 \sqrt{1-x^2}dx$ and click on Evaluate Numerically to obtain

$$\int_0^1 \sqrt{1-x^2}dx = .7854.$$

5. Point at the expression $\int_0^1 \sqrt{1-x^3}dx$ and click on Evaluate to obtain

$$\int_0^1 \sqrt{1-x^3}dx = \frac{2}{15}\left(\sqrt{\pi}\right)^3 \frac{\sqrt{3}}{\Gamma\left(\frac{2}{3}\right)\Gamma\left(\frac{5}{6}\right)}.$$

6. Point at the expression $\int_0^\infty x^3 e^{-x}dx$ and click on Calculus and then Integrate by Parts. You will see the dialog box

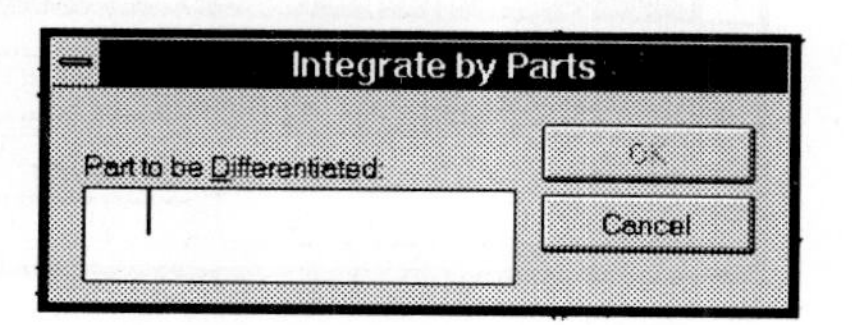

in which you should fill x^3 as the part to be differentiated. You will obtain

$$\int_0^\infty x^3 e^{-x}dx = -\int_0^\infty \left(-3x^2 e^{-x}\right) dx.$$

7. Point at the integral $\int_0^1 x^{a-1}(1-x)^{b-1} dx$ and click on Calculus and then Change Variable. You will see the change of variable dialog box.

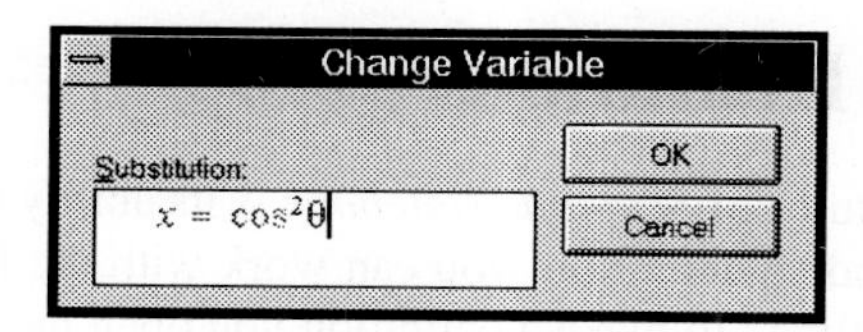

Type the command $x = \cos^2\theta$ as shown and click on OK. You will obtain

$$\int_0^{\frac{1}{2}\pi} 2\left(\cos^{(2a-1)}\theta\right)\left(1-\cos^2\theta\right)^{b-1}\sin\theta\, d\theta.$$

On your own initiative you might choose to rewrite this integral as

$$\int_0^{\frac{\pi}{2}} 2\sin^{2b-1}\theta\cos^{2a-1}\theta d\theta.$$

8. Point at the expression

$$\lim_{n\to\infty}\frac{(n!)\,e^n}{n^{n+\frac{1}{2}}}$$

and click on Simplify to obtain

$$\lim_{n\to\infty}\frac{(n!)\,e^n}{n^{n+\frac{1}{2}}}=\sqrt{2\pi}.$$

9. Point at the differential equation $(1-x^2)\,y''-xy'+y=x$ and click on Solve ODE and then Exact. A dialog box will come up asking you to give the name of the independent variable:

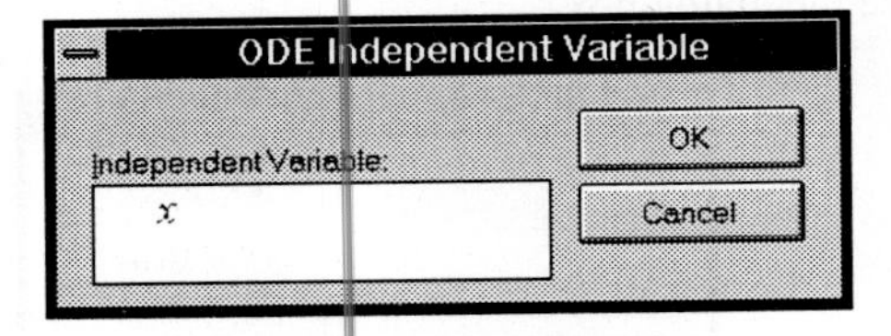

Type in x as shown and click on OK. You will obtain[3]

$$y(x)=-\frac{1}{2}\frac{-x\sqrt{(-1+x^2)}-\ln\left(x+\sqrt{(-1+x^2)}\right)+\left(\ln\left(x+\sqrt{(-1+x^2)}\right)\right)x^2}{\sqrt{(-1+x^2)}}+C_1x+C_2\sqrt{(-1$$

2.2 Making Definitions

One of the powerful features of *Scientific Notebook* is its ability to inform Maple of a definition. Once you have made a definition, you can work with the function by name instead of having to retype it each time. To make a definition you open the Maple menu and then click on Define. This brings up the define submenu.

[3] The next line can be seen completely in the on-screen version of this book.

New Definition
Undefine
Show Definitions...
Clear Definitions
Save Definitions
Restore Definitions
Define Maple Name...

If you point to an equation like $f(x) = \sqrt{1+x^2}$ and click on New Definition then you will inform *Scientific Notebook* that $f(x) = \sqrt{1+x^2}$ for all numbers x.

2.2.1 Some Examples of Definitions

1. Point at the equation $f(x) = x^3 - 2x^2 + 5x - 3$ and click on Define, and then on New Definition.

   (a) Type $f(1)$ and click on Evaluate to obtain $f(2) = 7$.

   (b) Type $f(x+3)$ and click on Expand to obtain

$$f(x+3) = x^3 + 7x^2 + 20x + 21$$

   (c) Type $f'(x)$ and click on Evaluate to obtain $f'(x) = 3x^2 - 4x + 5$.

   (d) Type $\frac{d}{dx} f(x)$ and click on Evaluate to obtain $\frac{d}{dx} f(x) = 3x^2 - 4x + 5$.

   (e) Type $f''(x)$ and click on Evaluate to obtain $f''(x) = 6x - 4$.

   (f) Type $\int_0^1 f(x)\,dx$ and click on Evaluate to obtain $\int_0^1 f(x)\,dx = -\frac{11}{12}$.

2. You can define a piecewise defined function by typing the separate parts into a matrix with three columns that is placed into a pair of brackets with a curly brace opening and an invisible close. For example, point at the equation

$$g(x) = \begin{cases} x+2 & \text{if} & x<0 \\ 2 & \text{if} & 0 \le x \le 1 \\ 1 & \text{if} & 1 < x \end{cases}$$

   and click on click on Define, and then on New Definition.

   (a) Type $g(-3)$ and click Evaluate to obtain $g(-3) = -1$.

   (b) Point at the equation $f(x) = x^2 - x - 1$ and click on Define, and then on New Definition. Then type $g(f(2))$ and click on Evaluate to obtain $g(f(2)) = 2$.

3. You can make one definition refer to another. Point at the equation $f(x) = x^2$ and click on click on Define, and then on New Definition. Then point at the equation $g(x) = (1 + f(x))^3$ and click on Define, and on New Definition. Now type $g(x)$ and click on Evaluate to obtain $g(x) = \left(1 + x^2\right)^3$.

4. Point at the equation $a_n = \frac{4^n (n!)^2}{(2n)!}$ and click on Define and on New Definition. You will see the Interpret Subscript dialog box

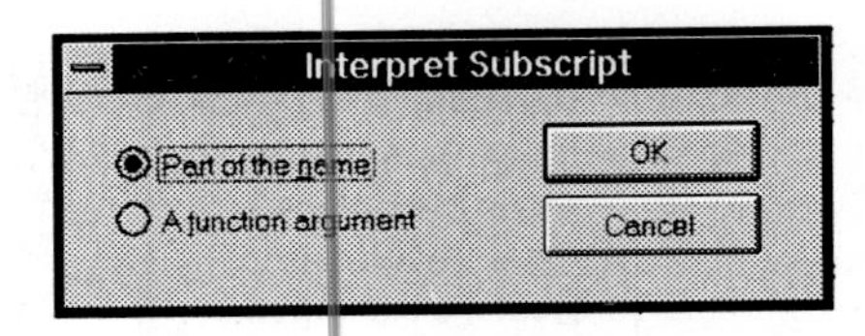

which asks you to choose whether the subscript n in the expression a_n is a function argument or whether it is just part of the name of a single symbol. Choose Function Argument. Now point at the expression $\frac{a_{n+1}}{a_n}$ and click on Simplify to obtain $\frac{a_{n+1}}{a_n} = 2\frac{n+1}{2n+1}$.

2.3 Evaluating a Function at a Column of Numbers

If f is a function of a real variable and A is a column of numbers of the form

$$A = \begin{bmatrix} a_1 \\ a_2 \\ \vdots \\ a_n \end{bmatrix}$$

then $f(A)$ is also a column of numbers. We define

$$f(A) = \begin{bmatrix} f(a_1) \\ f(a_2) \\ \vdots \\ f(a_n) \end{bmatrix}.$$

For example, point at the equation $f(x) = \frac{\sin x}{\sqrt{1+x^2}}$ and click on Define and New Definition and then point at the expression

$$f\begin{bmatrix} -85 \\ 50 \\ 57 \\ 92 \\ 54 \end{bmatrix}$$

and click on Evaluate Numerically. You will obtain

$$f\begin{bmatrix} -85 \\ 50 \\ 57 \\ 92 \\ 54 \end{bmatrix} = \begin{bmatrix} 2.0713 \times 10^{-3} \\ -5.2464 \times 10^{-3} \\ 7.6508 \times 10^{-3} \\ -8.472 \times 10^{-3} \\ -1.0346 \times 10^{-2} \end{bmatrix}$$

2.4 Iteration of Functions

Once a function f has been defined, it is easy to iterate the function starting at a given number. By the **iterations** of a function f starting at a number c we mean the numbers c, $f(c)$, $f(f(c))$, $f(f(f(c)))$ etc. The following exercises show how iterations may be obtained:

1. Point at the equation $f(x) = x^2$ and make this a new definition. Then open the Maple menu, click on Calculus and then click on Iterate. This brings up the iterate dialog box.

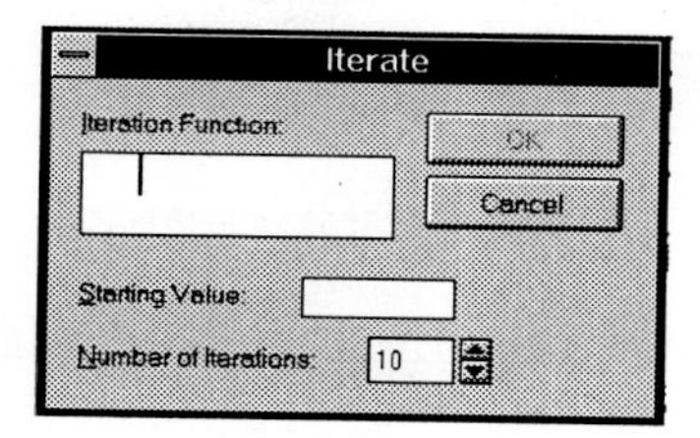

Type in f for the iteration function, 2 for the starting value and 6 for the number of iterations to obtain the iterates:

$$\begin{array}{c} 2 \\ 4 \\ 16 \\ 256 \\ 65536 \\ 42949\,67296 \\ 18446\,74407\,37095\,51616 \end{array}$$

2. Point at the equation $f(x) = x^2$ and make this a new definition. Now make the definition

$$g(x) = \begin{cases} \frac{f(x)}{2} & \text{if} \quad f(x) > 10 \\ 1 + f(x) & \text{if} \quad f(x) \leq 10 \end{cases}.$$

Now click on Calculus and Iterate in order to iterate the function g a total of eight times starting at the number 1. You will obtain the following iterates:

$$1$$

$$2$$

$$5$$

$$\frac{25}{2}$$

$$\frac{625}{8}$$

$$\frac{390625}{128}$$

$$\frac{15\,25878\,90625}{32768}$$

$$\frac{232\,83064\,36538\,69628\,90625}{21474\,83648}$$

$$\frac{54210\,10862\,42752\,21700\,37264\,00434\,97085\,57128\,90625}{9223\,37203\,68547\,75808}$$

2.5 Graphs in *Scientific Notebook*

In this chapter we shall explore some of the many kinds of graphs that can be drawn in a *Scientific Notebook* document. Don't forget that the figures in this book are not actual graphs. In order to learn the material properly you should redraw each graph as you come to it.

2.5.1 Drawing and Revising Graphs

We shall begin with the simplest type of graph; a two-dimensional graph drawn with rectangular coordinates. To illustrate the process we shall look at the graph of the equation $y = 8x^3 - 6x - 1$. Point at the expression $8x^3 - 6x - 1$ and click on Plot 2D. This opens the plot 2D submenu

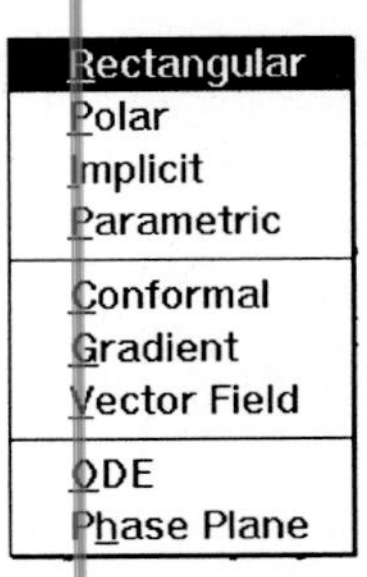

from which you can choose the type of 2D graph that you want to draw. Click on Rectangular. You will obtain the graph as follows: $8x^3 - 6x - 1$

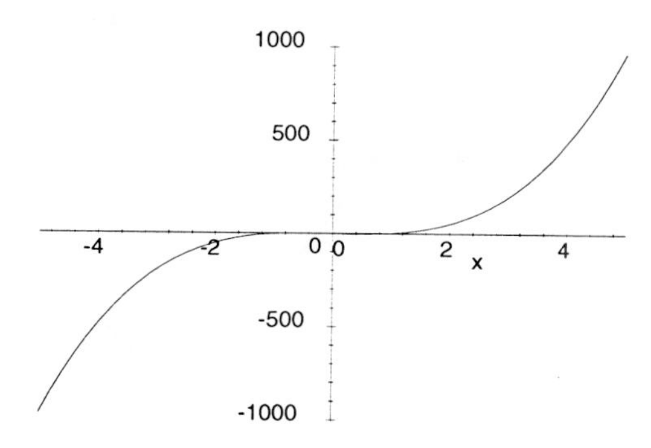

Now point at this figure and click. The figure will now have handles and a little box in its lower right corner.

Click on the little box at the lower right corner to bring up the Plot Properties dialog box.

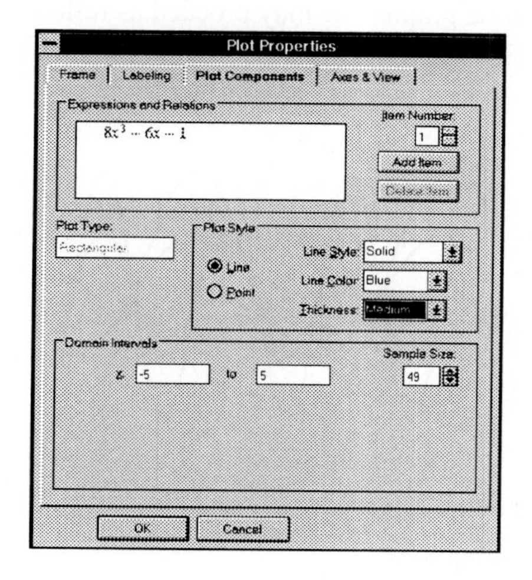

Click on Plot Components and choose the color and thickness of the plot. In the next figure we have chosen medium thickness and the color blue for the graph. After you click on OK, double-click the figure. At its right top corner you will see the following buttons:

Click on the middle one; the mountain to engage the zooming facility. If you now select any part of the graph with the mouse you will obtain a new figure that zooms onto the selected region. By zooming into the present graph you can make it like the graph in the figure shown.

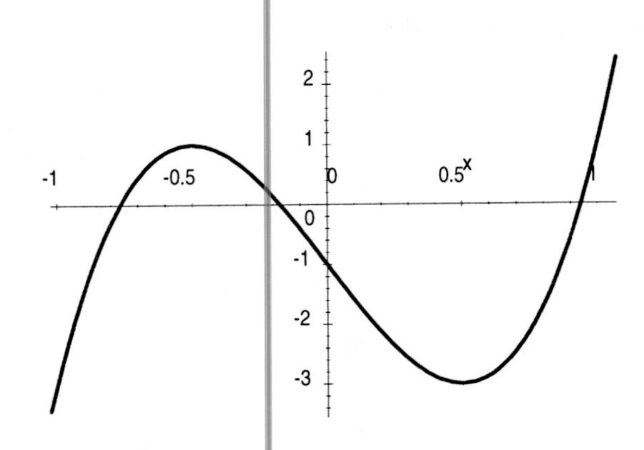

2.5.2 Multiple Rectangular Plots

The following figure shows the graphs of $\sin x$, $\sin 2x$, $2\sin x$ and $2\sin 2x$ on the same system of axes. To draw these graphs one may proceed as follows:

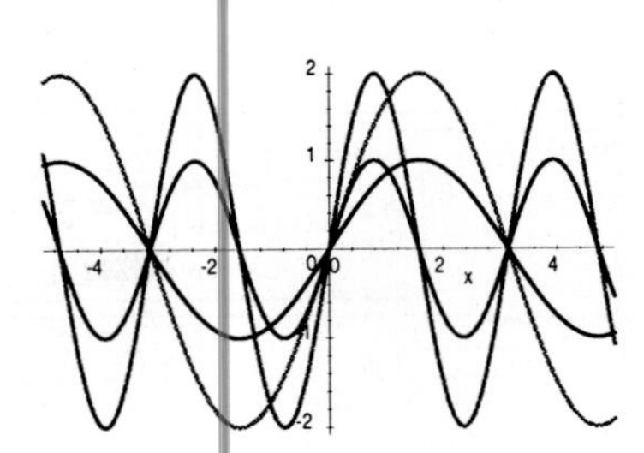

Point at the expression $\sin x$ and select Plot 2D and Rectangular (or simply click the button  that is in your computing toolbar).

- Highlight each of the other three expressions and drag them into the figure while holding down one of the mouse buttons.
- Select the figure and press control + F5 (or click the gray box at the right bottom of your figure) to revise the figure and change the colors of the individual plots according to your taste.

2.5.3 Implicit Plots

Graphs of equations of the type $f(x, y) = 0$ can be drawn by clicking on Plot 2D and then Implicit. The following two exercises illustrate this procedure.

2.5.3.1 Some Conic Sections

The conic sections $x^2 + y^2 = 1$, $x^2 - y^2 = 1$ and $x + y^2 = 0$ can all be plotted on the same coordinate axes as in the following figure. Highlight the first equation, open the Maple menu and click on Plot 2D and Implicit. Then highlight each of the other equations and drag it into your sketch. You will obtain the following figure:

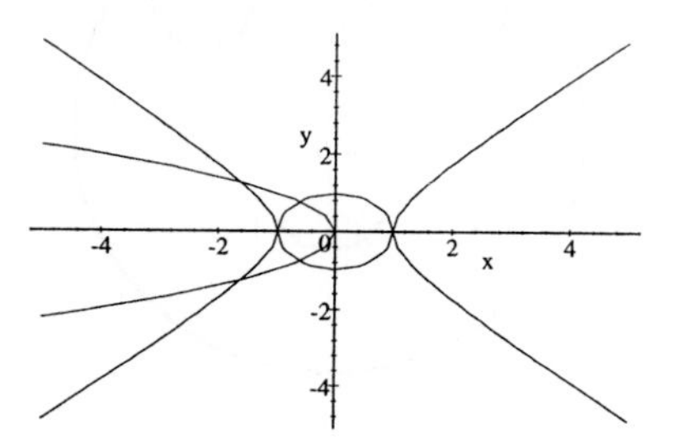

Zoom into this figure to restrict x to the interval $[-2, 2]$. Then open its plot properties dialog box, turn to Axes & View and check Equal Scaling Along Each Axis. Then turn to Plot Components and set your colors and set the plots to medium thickness. If necessary, increase the number of Sample Points. You should be able to make the figure come out as follows:

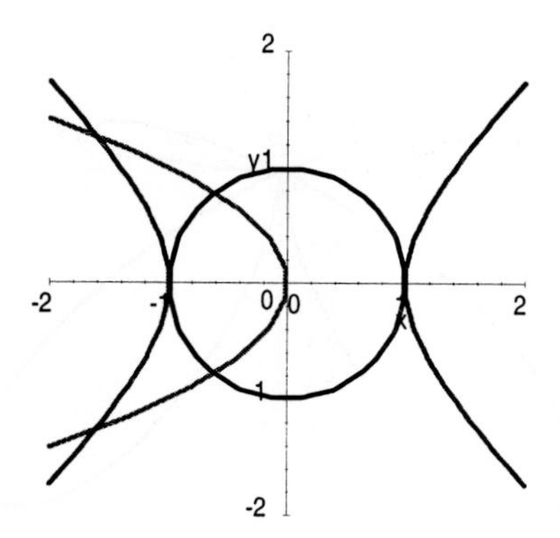

2.5.3.2 A Smiling Face

To obtain the smiling face shown in the next figure, draw a single sketch that contains the graphs

$$\begin{aligned} x^2 + y^2 &= 4 \\ (x+1)^2 + (y-1)^2 &= (0.4)^2 \\ (x-1)^2 + (y-1)^2 &= (0.4)^2 \end{aligned}$$

$$y - \frac{-1}{2} + \sqrt{\frac{1}{4} - x^2} = 0.$$

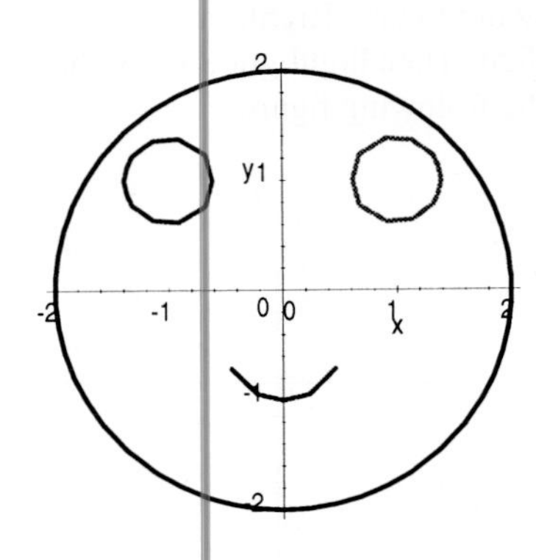

2.5.4 Polar Plots

2.5.4.1 A Simple Polar Plot

To draw the polar plot $r = \sin 2\theta \cos 3\theta$, point at the equation and click on Plot 2D and then Polar. You will obtain:

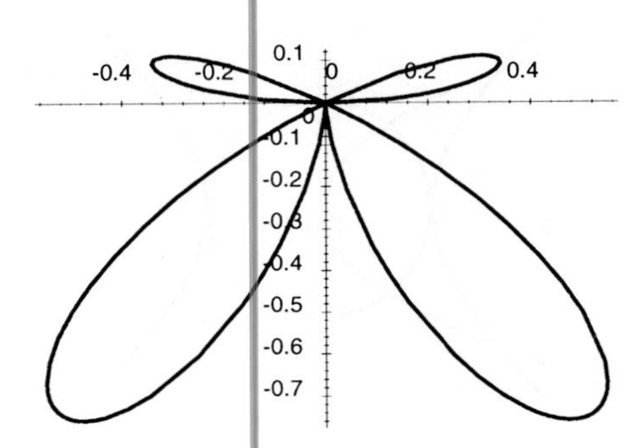

2.5.4.2 A Complicated Polar Plot

In this exercise we make a polar plot of the graph

$$r = 2 - \cos(7\theta) - \cos\left(\frac{32}{31}7\theta\right).$$

Point at this equation and click on Plot 2D and Polar. You will see the graph shown in the following figure.

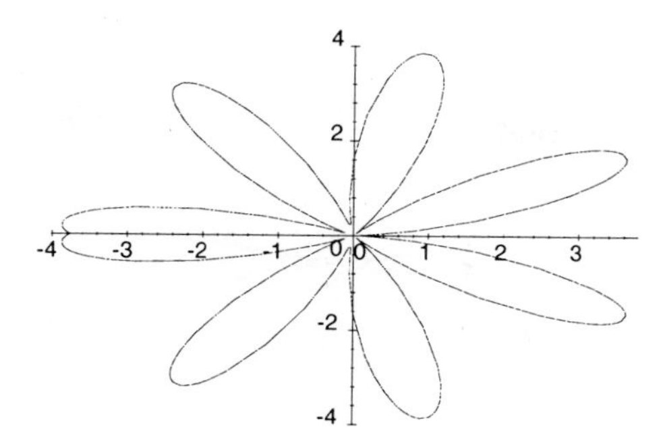

Now open the plot properties dialog box and Plot Components to set the domain interval to be $[0, 201]$. (Note that $64\pi \approx 201$.) You will obtain the graph shown in the following figure:

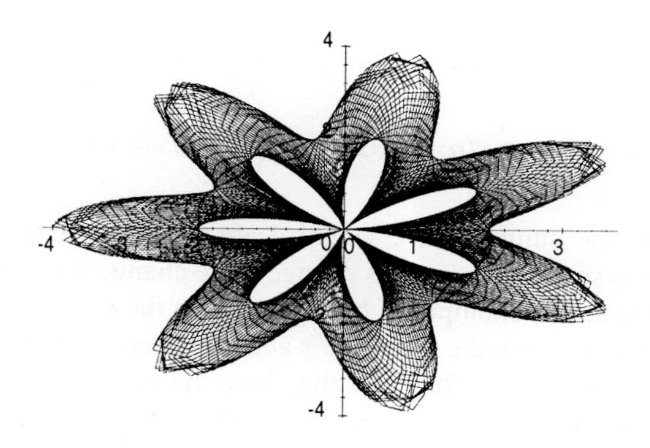

2.5.5 Parametric Plots

To draw the parametric curve

$$\begin{aligned} x &= t\cos 5t \\ y &= t\sin 5t - t \end{aligned}$$

point at the expression $(t\cos 5t, t\sin 5t - t)$ and click on Plot 2D and then Parametric Plot. The curve in the next figure shows the part of this graph as t varies from -20 to 20.

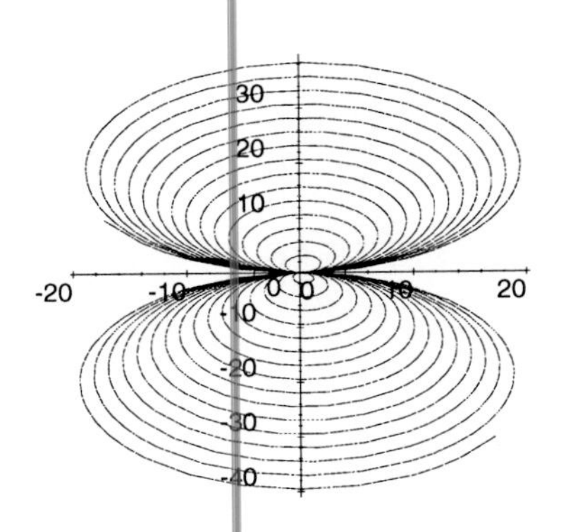

2.5.6 Exploring a Parametric Curve

In this section we show how *Scientific Notebook* can be used to explore the family of parametric curves

$$x = (a + b\cos t)\cos\left(\frac{t}{c} + d\cos t\right)$$
$$y = (a + b\cos t)\sin\left(\frac{t}{c} + d\cos t\right)$$

where, in each curve, the numbers a, b, c and d are given constants and $t \in [-20, 20]$. Specifically, we want to see how the curves change if we change the values of a, b, c and d.

Some things are clear. For example, if $b = d = 0$ then the curve is obviously a circle. If b is now allowed to increase then the values of x and y become larger when t is close to zero and smaller when t is close to π. What does this do to the graph? And what is the effect on the graph as c and d are changed? In order to answer these questions make the definitions

$$f(a, b, c, d, t) = (a + b\cos t)\cos\left(\frac{t}{c} + d\cos t\right)$$

and

$$g(a, b, c, d, t) = (a + b\cos t)\sin\left(\frac{t}{c} + d\cos t\right)$$

and then make parametric plots of the function

$$(f(a, b, c, d, t), g(a, b, c, d, t))$$

as t varies from -20 to 20 for some selected values of a, b, c and d. Finally, we have provided an animated sequence of graphs of this type that will help you in your explorations. This animation consists of six separate parts.

2.5.6.1 The First Part

We start off with

$$\begin{aligned} a &= 0 \\ b &= 0 \\ c &= 1 \\ d &= 0 \end{aligned}$$

and we let a increase from 0 to 2, in increments of 0.1.

2.5.6.2 The Second Part

We start off with

$$\begin{aligned} a &= 2 \\ b &= 0 \\ c &= 1 \\ d &= 0 \end{aligned}$$

and we let b increase from 0 to 1, in increments of 0.1.

2.5.6.3 The Third Part

This part is longer than the first two. We start off with

$$\begin{aligned} a &= 2 \\ b &= 1 \\ c &= 1 \\ d &= 0 \end{aligned}$$

and we let c increase from 1 to 5, in increments of 0.025. You will notice that, in spite of the very small size of this increment, there are places where you will wish that it had been even smaller. This is true especially at the beginning of this part of the animation.

2.5.6.4 The Fourth Part

In this part we start off with

$$\begin{aligned} a &= 2 \\ b &= 1 \\ c &= 5 \\ d &= 0 \end{aligned}$$

and we let d increase from 0 to 4, in increments of 0.2. This produces an interesting twisting effect on the graph.

2.5.6.5 The Fifth Part

In this part we start off with

$$\begin{aligned} a &= 2 \\ b &= 1 \\ c &= 5 \\ d &= 4 \end{aligned}$$

and we let a decrease from 2 to 1, in decrements of 0.1.

2.5.6.6 The Sixth Part

Finally we remove the twisting. We start off with

$$\begin{aligned} a &= 1 \\ b &= 1 \\ c &= 5 \\ d &= 4 \end{aligned}$$

and we let d decrease from 4 to 0, in decrements of 0.2.

To run the animation, hold down the control key and **click here.**

2.5.7 A Parametric Cone

In this subsection, we look at the parametric surface

$$\begin{aligned} x &= u\cos v \\ y &= u\sin v \\ z &= \frac{u}{2} \end{aligned}$$

Point at the expression $(u\cos v, u\sin v, u/2)$ and click on Plot 3D and then Rectangular. Open the Plot Properties dialog box for this cone and select Plot Components. Fill in the Domain Intervals in such a way that u runs from 0 to 2 and v runs from 0 to 6.283 (approximately 2π). Set the Plot Style to Patch, the Shading to XYZ and the Lighting to 0 as shown.

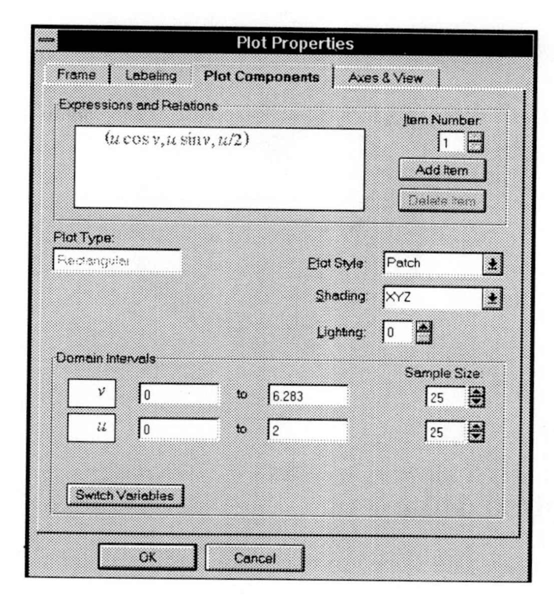

You will obtain the graph in the following figure:

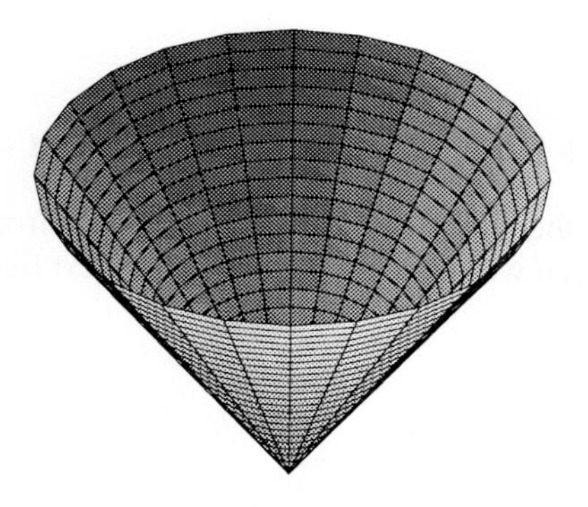

2.5.8 A Parametric Sphere

If we apply the method of the preceding subsection to the sphere

$$\begin{aligned} x &= \cos u \sin v \\ y &= \sin u \sin v \\ z &= \cos v \end{aligned}$$

we obtain the Maple plot in the next figure.

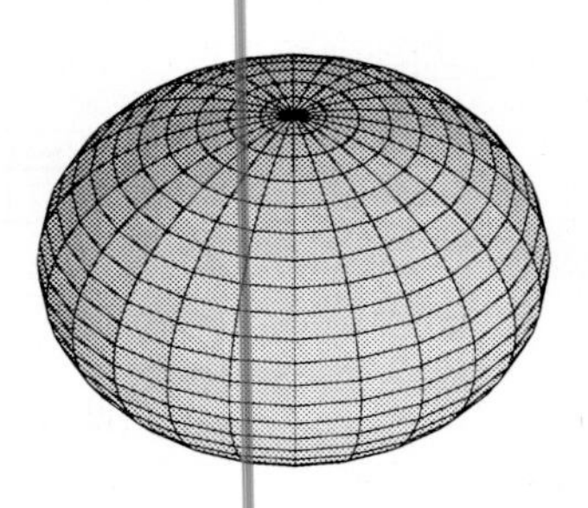

2.5.9 A Knotted Tube

In this subsection we draw a thickened form of the parametric graph

$$\begin{aligned} x &= -10\cos t - 2\cos(5t) + 15\sin(2t) \\ y &= -15\cos(2t) + 10\sin t - 2\sin(5t) \\ z &= 10\cos(3t). \end{aligned}$$

Click on the expression

$$(-10\cos t - 2\cos(5t) + 15\sin(2t), -15\cos(2t) + 10\sin t - 2\sin(5t), 10\cos(3t))$$

and click on Plot 3D and then Tube. If you allow t to vary from 0 to 6.3 and choose the radius of the tube to be 2, you will obtain the graph in the next figure.

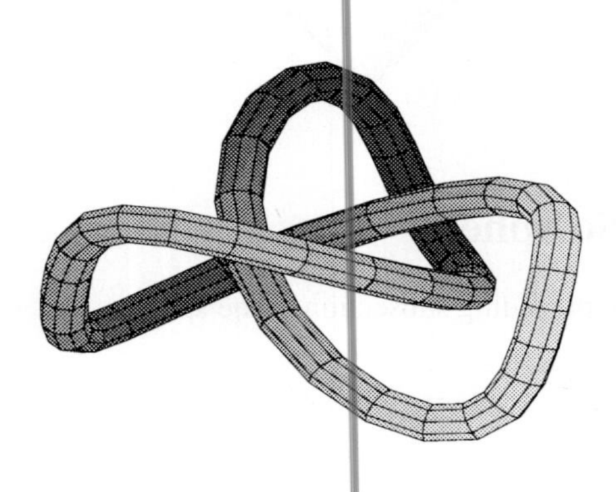

2.5.10 Another Tube

Try this one as an exercise. Make a tube plot of $\left(\cos^3 t, \sin^3 t, \sin(t + \frac{\pi}{4})\right) xy \sin x$ with radius .2 and letting t vary between 0 and 2π.

2.5.11 A Möbius Band

In this subsection we look at the Möbius band

$$\begin{aligned} x &= (1 - t\sin\theta)\cos 2\theta \\ y &= (1 - t\sin\theta)\sin 2\theta \\ z &= t\cos\theta \end{aligned}$$

where $-.4 \leq t \leq .4$ and $0 \leq \theta \leq \pi$. By pointing at the expression

$$((1 - t\sin\theta)\cos 2\theta, (1 - t\sin\theta)\sin 2\theta, t\cos\theta)$$

and clicking on Plot 3D and then Rectangular, and then setting the domain correctly, one obtains the Möbius band shown in the next figure.

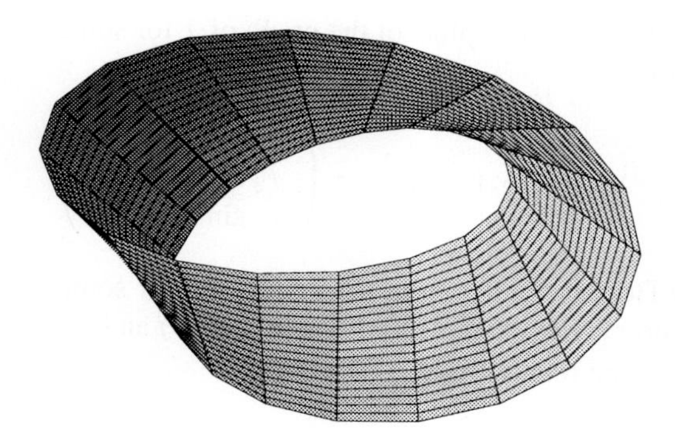

To see an animated view of this Möbius band, **click here.**

2.6 Some Miscellaneous Exercises

1. Draw the graph $y = x + \sin(2\pi x)$ and use the zooming feature of *Scientific Notebook* to study the behavior of the curve. Repeat this exercise for the graph $y = x^2 - 20\cos\left(x^2\right)$.
2. Draw the graph $z = \sin(x - \cos y)$. Rotate the surface to optimize your view of it.

3. Make the definition $f(x,a) = a(1-\cos x)$. Then assign some specific values to the number a and compare the graphs of f that are obtained by using Plot 2D Rectangular and Plot 2D Polar. Make t run from 0 to 2π.
4. Make the definition $f(t,a) = e^{at}$ and then produce some polar plots of the graph of f for some specific values of a. Repeat this exercise for $f(t,a) = \frac{a}{t}$.
5. Produce some polar plots of the function $f(t,a) = a\cos 3t$ and $f(t,a) = a\cos 2t$.
6. Make the definition

$$f(a,t) = \begin{pmatrix} \frac{3at}{1+t^3} \\ \frac{3at^2}{1+t^3} \end{pmatrix}$$

and produce the 2D parametric plots of the graph of f for some specific values of a.

7. Make the definition

$$f(a,t) = \begin{pmatrix} a\cos^3 t \\ a\sin^3 t \end{pmatrix}.$$

and produce the 2D parametric plots of the graph of f for some specific values of a.

8. Make the definition

$$f(a,t) = \begin{pmatrix} a(t-\sin t) \\ a(1-\cos t) \end{pmatrix}$$

and produce the 2D parametric plots of the graph of f for some specific values of a.

9. Make the definition

$$f(r_1,r_2,r_3,u,w) = \begin{pmatrix} r_1 \sin w \cos u \\ r_2 \cos w \cos u \\ r_3 \sin u \end{pmatrix}$$

and use Plot 3D Rectangular to plot the graph of f for some specific values of the numbers r_1, r_2 and r_3. For example, try $f(1,2,3,u,w)$ and $f(2,4,6,u,w)$.

10. Draw the graphs of the function

$$f(r,u,w) = \begin{pmatrix} ru\sin w \\ ru\cos w \\ r\sin u \end{pmatrix}$$

for some specific values of r.

11. Draw the graphs of the function

$$f(r,u,w) = \begin{pmatrix} r\cos w \\ r\sin w \\ u \end{pmatrix}$$

for some specific values of r.

12. Draw the graphs of the function

$$f(a, b, c, u, w) = \begin{pmatrix} a \sec u \sin w \\ b \sec u \cos w \\ cu \end{pmatrix}$$

for some specific values of r.

13. Draw the graph of the function

$$f(u, w) = \begin{pmatrix} u^2 \cos w \\ u \sin w \\ u \end{pmatrix}.$$

14. Draw the graphs of the function

$$f(r, u, w) = \begin{pmatrix} (r + \cos u) \cos w \\ (r + \sin w) \sin w \\ \sin u \end{pmatrix}$$

for some specific values of r.

Chapter 3
A Graphical View of Limits

In this chapter we shall demonstrate how *Scientific Notebook* can be used to obtain a graphical comparison between the rates at which two given functions approach limits. As we know,

$$\lim_{x\to 0}\frac{\sin x}{x}=1.$$

We shall use a graphical method to find a number k such that $\frac{\sin x}{x}$ approaches 1 at the same rate as $x^k \to 0$ when $x \to 0$. More precisely, we shall find a number k such that the limit

$$\lim_{x\to 0}\frac{1-\frac{\sin x}{x}}{|x|^k}$$

is finite and positive. We begin by defining

$$f(x,k)=\frac{1-\frac{\sin x}{x}}{|x|^k}$$

and we inform *Scientific Notebook* of this definition by clicking on Define and New Definition. By looking at the graphs of this function for some specific values of k, we can observe that whenever $k<2$ we have

$$\lim_{x\to 0} f(x,k)=0$$

and whenever $k>2$ we have

$$\lim_{x\to 0} f(x,k)=\infty.$$

When $k=2$, the limit is nonzero and finite. In fact, as you can see easily,

$$\lim_{x\to 0} f(x,2)=\frac{1}{6}.$$

We shall look at the graphs in two ways. First we shall look at an animated sequence of graphs of the type $y=f(x,k)$ for some values of k that increase from 1 to 3. To see this animation, **click here**. Observe the dramatic change that takes place in the graph as k increases through the value 2.

The second way in which we look at the graphs is to look at the surface $z = f(x, k)$ for $x \in [-1, 1]$ and $k \in [1.9, 2.1]$. See the next figure. Notice how the shape of this surface changes dramatically as k increases through the value 2.

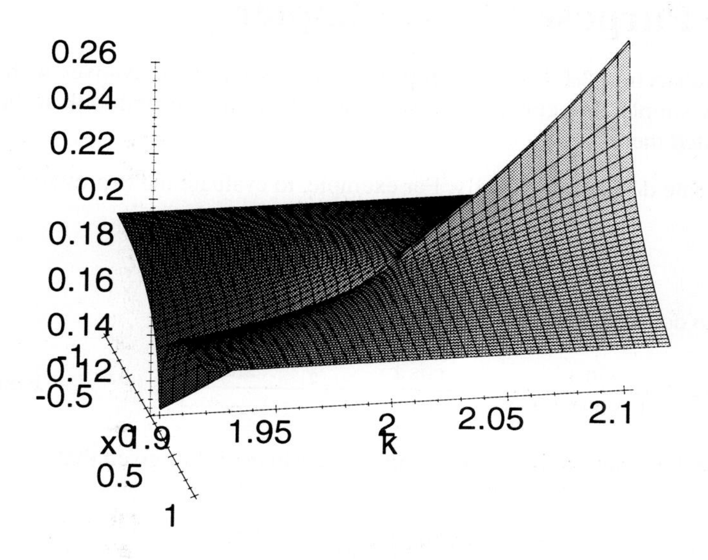

Chapter 4
Differential Calculus

4.1 The Purpose of this Chapter

As you saw in Subsection 2.1.4, the actual process of calculating derivatives with *Scientific Notebook* is very simple. For convenience we shall take a brief second look at the methods that were suggested there.

1. We can calculate derivatives directly. For example, to evaluate the expression

$$\frac{d}{dx}\frac{x^2 \sin x}{1+xe^{\cos x}},$$

all we have to do is point at it and click on Evaluate. We obtain

$$\frac{d}{dx}\frac{x^2 \sin x}{1+xe^{\cos x}} = 2x\frac{\sin x}{1+xe^{\cos x}} + x^2\frac{\cos x}{1+xe^{\cos x}} - x^2\frac{\sin x}{\left(1+xe^{\cos x}\right)^2}\left(e^{\cos x} - x\left(\sin x\right)e^{\cos x}\right).$$

2. We can name the function first. For example, to evaluate $f'(x)$ given that

$$f(x) = \frac{x^2 \sin x}{1+xe^{\cos x}},$$

all we have to do is point at the latter equation and click on Define and New Definition, and then point at $f'(x)$ and click on Evaluate. This yields

$$f'(x) = 2x\frac{\sin x}{1+xe^{\cos x}} + x^2\frac{\cos x}{1+xe^{\cos x}} - x^2\frac{\sin x}{\left(1+xe^{\cos x}\right)^2}\left(e^{\cos x} - x\left(\sin x\right)e^{\cos x}\right).$$

3. We can name the function and then evaluate the limit of its difference quotient. For example, to evaluate $f'(x)$ given that

$$f(x) = \frac{x^2 \sin x}{1+xe^{\cos x}},$$

all we have to do is point at the latter equation and click on Define and New Definition, and then point at

$$\lim_{t\to x}\frac{f(t)-f(x)}{t-x}$$

and click on Evaluate. This yields

$$\lim_{t \to x} \frac{f(t) - f(x)}{t - x} =$$

$$\frac{(\sin^2 x) x^3 e^{\cos x} + (\cos x) x^2 + x^2 (\sin x) e^{\cos x} + 2 (\sin x) x + (\cos x) x^3 e^{\cos x}}{1 + 2x e^{\cos x} + x^2 e^{2 \cos x}}.$$

In this chapter we are not merely concerned with the evaluation of derivatives. Instead, the chapter is designed to show how we may use *Scientific Notebook* to enhance our understanding of the *idea* of a derivative. Section 4.2 introduces three distinct ways in which we can approach the notion of a derivative and it shows how one may explore each of these approaches with the help of *Scientific Notebook*. We shall call these ways the **graphical** approach, the **numerical** approach and the **algebraic** approach to the definition.

Finally, in Section 4.3 we show how to use *Scientific Notebook* to make multiple plots that contain the graph of a given function together with the graphs of its derivative and second derivative. Such multiple plots can enhance your understanding of the role of the derivative and second derivative in describing the behavior of a given function.

4.2 Three Approaches to the Notion of a Derivative

4.2.1 The Graphical Approach

4.2.1.1 Introducing the Graphical Approach

Since the derivative of a function f at a number a in its domain is the limit

$$\lim_{h \to 0} \frac{f(a+h) - f(a)}{h},$$

the derivative can be illustrated graphically by drawing a family of lines through the point $(a, f(a))$ with slopes

$$\frac{f(a+h) - f(a)}{h}$$

for a variety of values of h that are taken closer and closer to 0. For each chosen value of h, the equation of the line that joins the points $(a, f(a))$ and $(a+h, f(a+h))$ is

$$y = f(a) + \left(\frac{f(a+h) - f(a)}{h} \right) (x - a).$$

We therefore proceed as follows:

1. We inform *Scientific Notebook* of the definition of the function f with which we are working.
2. We inform *Scientific Notebook* of the definition

$$L(x,a,h) = f(a) + \left(\frac{f(a+h) - f(a)}{h}\right)(x-a)$$

to enable us to draw the secant lines and we inform *Scientific Notebook* of the definition

$$T(x,a) = f(a) + f'(a)(x-a)$$

to enable us to draw the tangent line to the graph of f at the point $(a, f(a))$.

3. We draw the graphs of $L(x,a,h)$ for some chosen values of h and we draw the graph of $T(x,a)$.

4.2.1.2 Illustration of the Graphical Approach

In this example we shall illustrate the derivative of the function f defined by

$$f(x) = x^3 - 6x^2 - 5x + 1$$

at the point where $x = -1.5$. We point at each of the equations

$$a = -1.5$$

$$f(x) = x^3 - 6x^2 - 5x + 1$$

$$L(x,a,h) = f(a) + \left(\frac{f(a+h) - f(a)}{h}\right)(x-a)$$

$$T(x,a) = f(a) + f'(a)(x-a)$$

and click on Define and then New Definition. Now we point at the expression $f(x)$ and click on Plot 2D and Rectangular. After a bit of zooming we obtain the graph shown in the next figure.

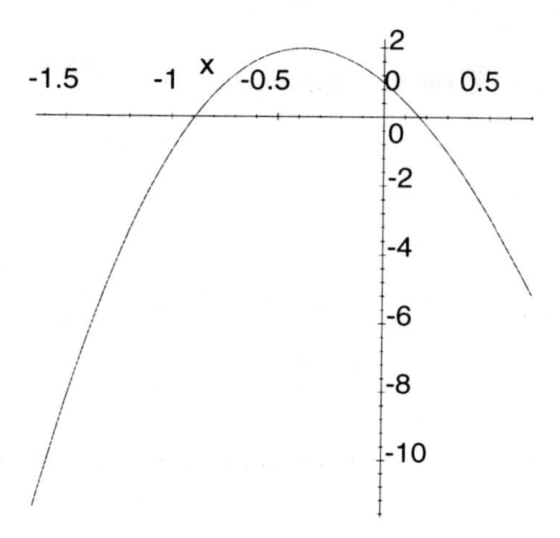

We now choose some values of h. We highlight each of the expressions $L(x, a, 2)$, $L(x, a, 1)$, $L(x, a, .5)$, $L(x, a, .2)$ $L(x, a, .1)$ in turn and drag it into the graph. Finally, we highlight the expression $T(x, a)$ and drag it into the graph. This gives us the next figure. For convenience we have shown the tangent line in red, the graph of f in blue and all the secant lines in black.

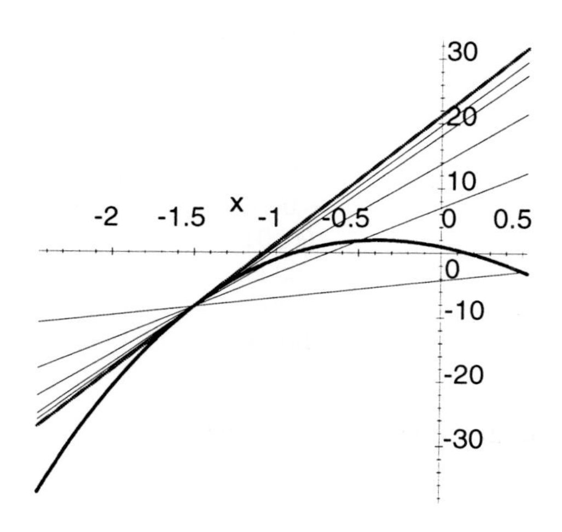

To see an animated view of the secant lines $y = L(x, a, h)$ as h decreases from 2 to 0.05, hold down your control button and .

4.2.2 The Numerical Approach

4.2.2.1 Introducing the Numerical Approach

Suppose that we want to compute the derivative of a given function f at a given number a. We need to find the limit

$$\lim_{h\to 0}\frac{f(a+h)-f(a)}{h}.$$

The numerical approach to the finding of $f'(a)$ is to evaluate the quotient

$$\frac{f(a+h)-f(a)}{h}$$

for a variety of small values of h. In order to evaluate $f'(a)$ numerically, we proceed as follows:

1. We inform *Scientific Notebook* of the definition of the function f with which we are working.
2. We inform *Scientific Notebook* of the value of the number a.
3. We inform *Scientific Notebook* of the definition

$$Q(h)=\frac{f(a+h)-f(a)}{h}.$$

4. We evaluate the function Q on the column of numbers

$$\begin{array}{c} .1 \\ -.1 \\ .01 \\ -.01 \\ .001 \\ -.001 \\ .0001 \\ -.0001 \\ .00001 \\ -.00001 \\ .000001 \\ -.000001 \end{array}$$

4.2.2.2 Illustration of the Numerical Approach

In this example we define

$$f(x)=x^2\cos\left(\sqrt{1+2^x}\right)$$

and we want to find $f'(3)$. We proceed as follows:

1. We point at the equation

$$f(x) = x^2 \cos\left(\sqrt{1+2^x}\right)$$

and click on Define and New Definition.

2. We point at $a = 3$ and click on Define and New Definition.

3. We point at

$$Q(h) = \frac{f(a+h) - f(a)}{h}$$

and click on Define and New Definition.

4. We point at the expression

$$Q\begin{pmatrix} .1 \\ -.1 \\ .01 \\ -.01 \\ .001 \\ -.001 \\ .0001 \\ -.0001 \\ .00001 \\ -.00001 \\ .000001 \\ -.000001 \end{pmatrix}$$

and click on Evaluate Numerically to obtain

$$Q\begin{pmatrix} .1 \\ -.1 \\ .01 \\ -.01 \\ .001 \\ -.001 \\ .0001 \\ -.0001 \\ .00001 \\ -.00001 \\ .000001 \\ -.000001 \end{pmatrix} = \begin{pmatrix} -6.8929 \\ -7.2574 \\ -7.0953 \\ -7.1314 \\ -7.1119 \\ -7.1156 \\ -7.1136 \\ -7.1139 \\ -7.1138 \\ -7.1138 \\ -7.114 \\ -7.113 \end{pmatrix}.$$

This suggests that the value of $f'(3)$ is approximately -7.1. In fact, if we point at $f'(3)$ and click on Evaluate Numerically we obtain $f'(3) = -7.113758$.

We emphasize that the method described in this example is meant as a teaching tool. It is not offered an efficient and reliable way of evaluating a derivative numerically. As a matter of fact, if you continue this method by taking h as small as 10^{-8}, or smaller, then Maple will not handle the difference quotient correctly and will give inaccurate results.

4.2.3 The Algebraic Approach

4.2.3.1 Introducing the Algebraic Approach

The algebraic approach is to use the symbolic capability of *Scientific Notebook* to simplify the difference quotient

$$\frac{f(a+h)-f(a)}{h}$$

so that it becomes easy to find its limit as $h \to 0$. Alternatively, it may be convenient to simplify the difference quotient in the form

$$\frac{f(t)-f(a)}{t-a}$$

so that it becomes easy to find its limit as $t \to a$.

4.2.3.2 Some Illustrations of the Algebraic Approach

1. We shall find $f'\left(\frac{3}{7}\right)$ given that $f(x) = -\frac{1}{x^4} + x^3 - 2x^2$ for each x. We begin by pointing at the equation $f(x) = -\frac{1}{x^4} + x^3 - 2x^2$ and clicking on Define and New Definition. Then we point at

$$\frac{f\left(\frac{3}{7}+h\right)-f\left(\frac{3}{7}\right)}{h}$$

and click on Simplify to obtain[4]

$$\frac{f\left(\frac{3}{7}+h\right)-f\left(\frac{3}{7}\right)}{h} =$$

$$= \frac{1}{3969}\frac{88568667 + 307579167h + 470204931h^2 + 258970831h^3 - 12252303h^4 + 9529569h^5 + 952}{(3+7h)^4}$$

Finally, we give *Scientific Notebook* the definition $h = 0$ and point at the expression

$$\frac{1}{3969}\frac{88568667 + 307579167h + 470204931h^2 + 258970831h^3 - 12252303h^4 + 9529569h^5 + 95295}{(3+7h)^4}$$

and click on Simplify again to obtain

$$\frac{1}{3969}\frac{88568667 + 307579167h + 470204931h^2 + 258970831h^3 - 12252303h^4 + 9529569h^5 + 95295}{(3+7h)^4}$$

2. In this example we show how to use *Scientific Notebook* to perform the algebraic manipulations that are required in order to evaluate the derivative of a function that

[4] The next few lines can be seen completely in the on-screen version of this book.

contains rational exponents. The key to this problem is the algebraic identity

$$t^{\frac{m}{n}} - x^{\frac{m}{n}} = \left(t^{\frac{1}{n}} - x^{\frac{1}{n}}\right) \sum_{j=1}^{m-1} t^{\frac{m-1-j}{n}} x^{\frac{j-1}{n}}$$

that holds whenever t and x are positive numbers and m and n are positive integers. With this identity in mind we point at the equation

$$g(t,x,m,n) = \sum_{j=0}^{m-1} t^{\frac{m-1-j}{n}} x^{\frac{j}{n}}$$

and click on Define and New Definition. We shall now obtain a formula for $f'(x)$ given that $f(x) = x^{\frac{4}{7}}$ for all $x > 0$. Given a positive number $t \neq x$ we have

$$\frac{f(t) - f(x)}{t - x} = \frac{t^{\frac{4}{7}} - x^{\frac{4}{7}}}{t^{\frac{7}{7}} - x^{\frac{7}{7}}} = \frac{\left(t^{\frac{1}{7}} - x^{\frac{1}{7}}\right) g(t,x,4,7)}{\left(t^{\frac{1}{7}} - x^{\frac{1}{7}}\right) g(t,x,7,7)}$$

$$= \frac{g(t,x,4,7)}{g(t,x,7,7)}.$$

Once we have cancelled the factor $t^{\frac{1}{7}} - x^{\frac{1}{7}}$ we can evaluate

$$\lim_{t \to x} \frac{g(t,x,4,7)}{g(t,x,7,7)}$$

simply by substituting $t = x$. We therefore point at the equation $t = x$ and click on Define and New Definition and then point at the expression

$$\frac{g(t,x,4,7)}{g(t,x,7,7)}$$

and click on Simplify yielding

$$\lim_{t \to x} \frac{g(t,x,4,7)}{g(t,x,7,7)} = \frac{4}{7\left(\sqrt[7]{x}\right)^3}.$$

4.2.4 Some Exercises

Give a graphical and a numerical discussion of each of the following derivatives:

1. $f'(0)$ given that $f(x) = 4x + 3x^{4/3}$ for each x.
2. $f(\pi/2)$ given that $f(x) = (2 + \sin x)^{\cos x}$.
3. $f'(0)$ given that $f(x) = 4x + 3x^{4/3}$ for each x.

4.3 Using Multiple Plots to Illustrate Derivatives

4.3.1 Introduction and Example

The graphing capability of *Scientific Notebook* makes it particularly easy to sketch the graph of a given function f alongside the graphs of its derivative f' and second derivative f''. In this section we show how this can be done and we provide an example that demonstrates how sketches of this sort can be used by students to deepen their appreciation of the role of the derivative and second derivative of a function.

We begin by defining

$$f(x) = \arctan\left(\frac{2x^2-1}{2x^2+1}\right).$$

In order to inform *Scientific Notebook* of this definition, we point at the equation and click on Define and New Definition. We now apply the following step by step procedure:

1. In order to illustrate the fact that *Scientific Notebook* recognizes the notion $f'(x)$ for the first derivative of f, we point at the expression $f'(x)$ and click on Evaluate to obtain

$$f'(x) = \frac{4\frac{x}{2x^2+1} - 4\frac{2x^2-1}{(2x^2+1)^2}x}{1+\frac{(2x^2-1)^2}{(2x^2+1)^2}}.$$

If we point at the latter expression and click on Simplify, we obtain

$$f'(x) = \frac{4\frac{x}{2x^2+1} - 4\frac{2x^2-1}{(2x^2+1)^2}x}{1+\frac{(2x^2-1)^2}{(2x^2+1)^2}} = \frac{4x}{4x^4+1}.$$

Similarly, by pointing at the expression $f''(x)$, evaluating and simplifying, we obtain

$$f''(x) = \frac{\frac{4}{2x^2+1} - 32\frac{x^2}{(2x^2+1)^2} + 32\frac{2x^2-1}{(2x^2+1)^3}x^2 - 4\frac{2x^2-1}{(2x^2+1)^2}}{1+\frac{(2x^2-1)^2}{(2x^2+1)^2}}$$

$$-\frac{4\frac{x}{2x^2+1} - 4\frac{2x^2-1}{(2x^2+1)^2}x}{\left(1+\frac{(2x^2-1)^2}{(2x^2+1)^2}\right)^2}\left(8\frac{2x^2-1}{(2x^2+1)^2}x - 8\frac{(2x^2-1)^2}{(2x^2+1)^3}x\right)$$

$$= -4\frac{12x^4-1}{(4x^4+1)^2}.$$

2. Now we plot the graphs of the functions f, f' and f''. To do this, we point at the expression $f(x)$, open the Maple menu and click on Plot 2D and Rectangular. Then we point at each of the expressions $f'(x)$ and $f''(x)$, highlight it and drag it into the

sketch. After opening the plot properties dialog box and changing the colors we can obtain the sketch in the form shown in the next figure:

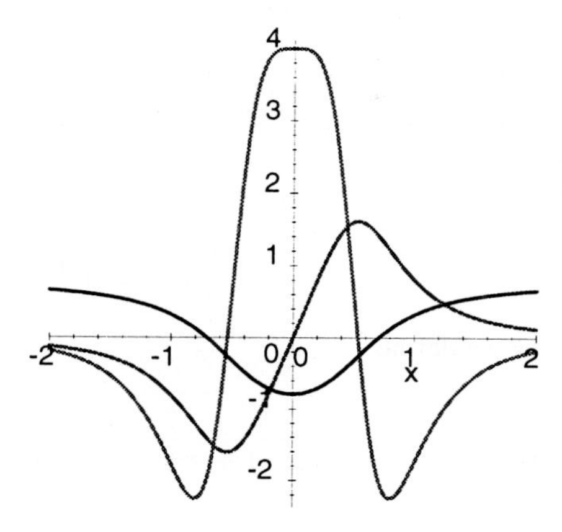

An instructive exercise is to determine from this figure which of the three curves is the graph of f, which is the graph of f' and which is the graph of f''. Explain carefully how you reach your conclusions.

4.3.2 Exercises

In each of the following exercises, repeat the procedure described above to analyze the graph of the given function f and the graphs of its first and second derivatives:

1. For every real number x we define

$$f(x) = \ln\left(1 + x^2\right).$$

2. For every real number x we define

$$f(x) = (2 + \sin x)^{\cos x}.$$

3. Whenever $\sin x \neq -1$ we define

$$f(x) = (1 + \sin x)^{\cos x}.$$

4. Whenever $\sin x \neq -1$ we define

$$f(x) = (1 + \sin x)^{|\sin x|}.$$

5. Whenever $\sin x \neq 1$ we define

$$f(x) = (1 - |\sin x|)^{|\tan x|}.$$

6. Whenever $\sin x \neq 0$ we define

$$f(x) = |\sin x|^{\sin x}.$$

7. For every real number x we define

$$f(x) = \frac{\ln(1+x^2)}{1+x+x^2}.$$

8. For every real number x we define

$$f(x) = \ln\left(1+\sin\left(\frac{1}{1+x^2}\right)\right).$$

9. For every real number x we define

$$f(x) = \frac{e^{\sqrt{\ln(1+x^2)}}}{1+x^2}.$$

10. Whenever $x \neq 0$ we define

$$f(x) = \frac{x^2}{2} + x^3 \sin\left(\frac{1}{x}\right).$$

Plot the graphs of f and f' only and examine the graph for values of x that are close to 0.

Chapter 5
Maxima and Minima

In this chapter we shall explore some ways in which *Scientific Notebook* can be used to find the maxima and minima of a given function. Our method will involve the following operations:

1. Sketching a graph in order to help us see the maximum and minima that we want to find.
2. Finding the critical points of the function.
3. Applying the second derivative test at the critical points.
4. Evaluating the function at its critical points.

5.1 Functions of One Variable

The role of the critical points in finding the maxima and minima of a function of one variable is described in the following elementary theorem that is sometimes called **Fermat's theorem.**

5.1.1 Fermat's Theorem

Suppose that a function f is defined on an open interval and has either a local maximum value or a local minimum value at a number c in that interval. Then either $f'(c) = 0$ or $f'(c)$ does not exist.

Once the critical points have been found, we can, if we wish, investigate the local behavior of the function at each critical point by means of the second derivative test:

5.1.2 The Second Derivative Test

Suppose that f is a differentiable function on an open interval, that c is a number in the interval and that $f'(c) = 0$.

1. *If $f''(c) > 0$ then the function f has a local minimum value at the number c.*
2. *If $f''(c) < 0$ then the function f has a local maximum value at the number c.*

5.1.3 Example

Given that

$$f(x) = \left(x^3 - x - 1\right) x^3 e^{-x^2}$$

for every real number x, find the local maximum and minimum points of f and find the maximum and minimum values of this function.

We begin by pointing at the equation that defines this function f and clicking on Define and New Definition. The next thing to do is to sketch the graph of this function. Point at the expression $f(x)$ and click on Plot 2D and Rectangular (or simply click on the button which can be found in your toolbars. The sketch will appear as shown in the next figure.

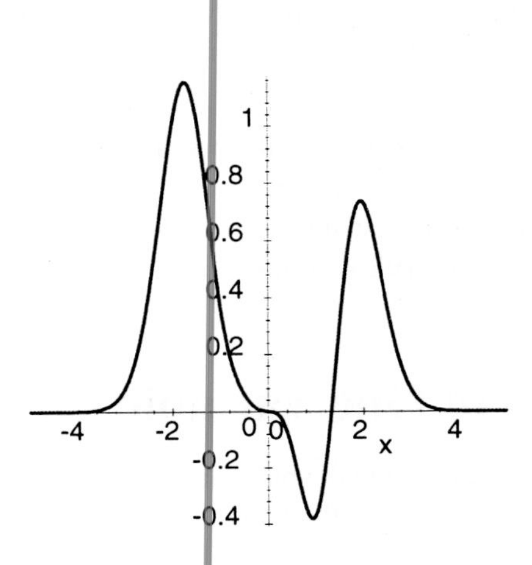

We can see that this function has a maximum at a number close to -2 and a minimum at a number close to 1. To find the critical points of f we point at the expression $f'(x)$ and click on Evaluate. Then highlight the expression obtained for $f'(x)$, hold down the control button, and click on Factor. We obtain

$$f'(x) = -x^2 e^{-x^2}\left(-8x^3 + 4x + 3 + 2x^5 - 2x^2\right)$$

and by pointing at the expression $-x^2\left(-8x^3 + 4x + 3 + 2x^5 - 2x^2\right)$ and clicking on Solve and Numeric we obtain the critical numbers as

$$\{x = 0\}, \{x = 0\}, \{x = -1.7476\}, \{x = .92589\}, \{x = 1.9446\}.$$

Now click on the button in your toolbar to make a matrix, select one column and five rows and drag in the critical numbers. You will have the matrix

$$\begin{matrix} 0 \\ 0 \\ -1.7476 \\ .92589 \\ 1.9446 \end{matrix}$$

Finally, evaluate the function f at this matrix. You will see

$$f\begin{pmatrix} 0 \\ 0 \\ -1.7476 \\ .92589 \\ 1.9446 \end{pmatrix} = \begin{pmatrix} 0 \\ 0 \\ 1.1554 \\ -.38131 \\ .73883 \end{pmatrix}$$

from which you can see that the function f has a maximum value of about 1.1554 at the number -1.7476 and has a minimum of about $-.38131$ at the number .92589.

5.1.4 Some Exercises

In each of the following exercises, find the maximum and the minimum values of the given function.

1. For every $x \in [-4, 4]$ we define
$$f(x) = x\cos x + x^2 \cos 2x.$$

2. For every real number x we define
$$f(x) = \frac{x\cos x + x^2\cos 2x}{1 + x^4}.$$

3. For every $x \in [-5, 5]$ we define
$$f(x) = \cos(x \sin x).$$

4. For every real number x we define
$$f(x) = \frac{x + \cos(x\sin x)}{1 + x^2}.$$

5. For every $x \in [0, 12]$ we define
$$f(x) = \left(\frac{(x + x^2 |\sin x|)}{(x^2 + 1)}\right)^{1/3}.$$

5.1.5 The Ladder in the Corridor Problem

This problem appears in the *Maple V Flight Manual, Tutorials for Calculus, Linear Algebra, and Differential Equations* where it is solved using commands that are written in Maple syntax. We shall use *Scientific Notebook* to call up those Maple operations without writing any Maple syntax.

5.1.5.1 Statement of the Problem

We wish to determine the longest ladder that can fit around a right angle turn from a hallway of width b into a passageway of width a. The ladder must be held horizontal. The next figure illustrates the given information.

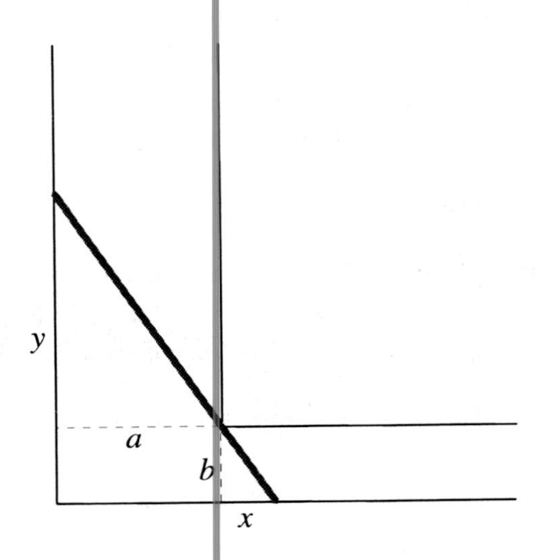

5.1.5.2 Solution of the Problem

First we observe that if the length of the ladder is l then

$$l = \sqrt{(x+a)^2 + (y+b)^2}.$$

We see from the similarity of the two triangles in the figure that

$$\frac{y}{b} = \frac{a}{x}$$

and so

$$y = \frac{ab}{x}.$$

In order to eliminate y from the expression for l, we point at the equation

$$y = \frac{ab}{x}$$

and click on Define and New Definition. Then we point at the expression

$$\sqrt{(x+a)^2 + (y+b)^2}$$

and click on Simplify yielding

$$l = \frac{x+a}{x}\sqrt{(x^2+b^2)}.$$

Thus the problem boils down to the problem of finding the minimum value of the function f defined by

$$f(x) = \frac{x+a}{x}\sqrt{(x^2+b^2)}$$

for $x > 0$. We observe that $f(x) \to \infty$ as $x \to 0_+$ and that $f(x) \to \infty$ as $x \to \infty$. Therefore f must have a minimum value at a value of x for which $f'(x) = 0$. We now point at the definition of $f(x)$ and click on Define and New Definition. Then we point at the equation $f'(x) = 0$ and click on Solve and Exact. This brings up the dialog box

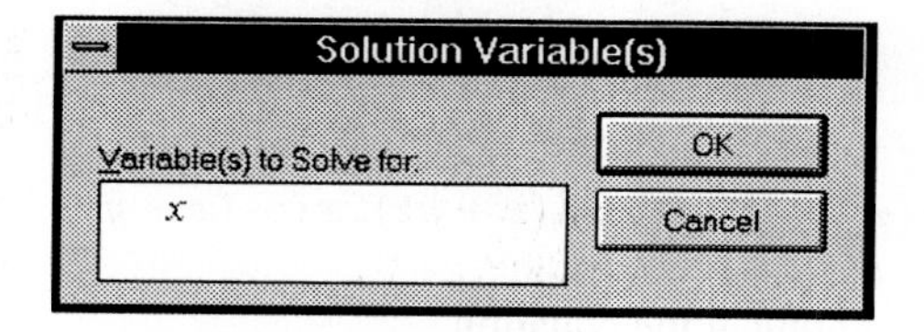

and we fill in x as shown and click on OK. This gives us:

$$\text{Solution is : } \{x = \rho\} \text{ where } \rho \text{ is a root of } -ab^2 + Z^3$$

which means that $x = \sqrt[3]{ab^2}$. Finally, we point at the equation $x = \sqrt[3]{ab^2}$ and click on Define and New Definition and then we point again at $f(x)$ and click on Simplify. We obtain

$$f(x) = \frac{\sqrt[3]{a}\left(\sqrt[3]{b}\right)^2 + a}{\sqrt[3]{a}\left(\sqrt[3]{b}\right)^2}\sqrt{\left(\left(\sqrt[3]{a}\right)^2\left(\sqrt[3]{b}\right)^4 + b^2\right)}.$$

We see that Maple hasn't done such a good job of simplifying this expression. We could have done better ourselves. But this doesn't matter. The latter expression is the length of the longest possible ladder that will go around the corner for given values of a and b. For example if we are told that $a = 3$ and $b = 4$ then we can point to each of these equations and click on Define and New Definition and then point to $f(x)$ and click on Evaluate Numerically to yield

$$f(x) = 9.865663.$$

5.2 Functions of Several Variables

5.2.1 Fermat's Theorem for Functions of Several Variables

The analogue of Fermat's theorem for a function of more than one variable refers to the **gradient** of the function. The gradient of a function f is written as ∇f using the symbol ∇ that you can find in your toolbars, or which you can produce by holding down your control key and typing the name *nabla* of this symbol. If f is a function defined on a region of $\mathbf{R}^n$ then the gradient ∇f at a point $(x_1, \cdots, x_n)$ is defined by the equation

$$\nabla f(x_1, \cdots, x_n) = \left(\frac{\partial f}{\partial x_1}, \frac{\partial f}{\partial x_2}, \cdots, \frac{\partial f}{\partial x_n}\right).$$

For example, if

$$f(x, y, z) = xy\cos(x + yz)$$

then[5]

$$\nabla f(x, y, z) = \left(y\cos(x + yz) - xy\sin(x + yz), x\cos(x + yz) - xy(\sin(x + yz))z, -xy^2\sin(x +\right.$$

which you can see by pointing at the equation

$$f(x, y, z) = xy\cos\left(x + yz^3\right)$$

and clicking on Define and New Definition and then pointing at the expression $\nabla f(x, y, z)$ and clicking on Evaluate. Using this notation we can state the n-dimensional version of Fermat's theorem as follows:

Suppose that a function f is defined on an open region in $\mathbf{R}^n$ and has either a local maximum value or a local minimum value at a **c** *in that region. Then either $\nabla f(x_1, \cdots, x_n)$ is the zero vector or $\nabla f(x_1, \cdots, x_n)$ does not exist.*

The points $(x_1, \cdots, x_n)$ in the region at which either $\nabla f(x_1, \cdots, x_n)$ is the zero vector or $\nabla f(x_1, \cdots, x_n)$ does not exist are called the **critical points** of the function f.

5.2.2 Finding the Critical Points of a Function

In order to find the critical points of a function f at a point $(x_1, \cdots, x_n)$ we need to solve the equation $\nabla f(x_1, \cdots, x_n) = \mathbf{0}$ where $\mathbf{0}$ is the zero vector in $\mathbf{R}^n$. To solve this equation using *Scientific Notebook* we click on the button [matrix button] in the toolbar to insert a matrix and

---

[5] The next line can be seen completely in the on-screen version of this book.

select one column and n rows. Then we fill in the equations as

$$\frac{\partial f(x_1,\cdots,x_n)}{\partial x_1} = 0$$

$$\frac{\partial f(x_1,\cdots,x_n)}{\partial x_2} = 0$$

$$\vdots$$

$$\frac{\partial f(x_1,\cdots,x_n)}{\partial x_n} = 0$$

and we click on Solve and Exact or we click on Solve and Numeric. The following examples illustrate this procedure.

5.2.3 Finding Critical Points Exactly

1. Find the critical points of the function f defined by the equation

$$f(x, y, z) = z^2 + x + xy \cos(x + yz)$$

for every point $(x, y, z) \in \mathbf{R}^3$. After supplying the definition of this function to *Scientific Notebook* we point at the system of equations

$$\frac{\partial f(x,y,z)}{\partial x} = 0$$

$$\frac{\partial f(x,y,z)}{\partial y} = 0$$

$$\frac{\partial f(x,y,z)}{\partial z} = 0$$

and we click on Solve and Exact. We obtain: Solution is : $\{z = 0, x = 0, y = -1\}$. The critical point of f is therefore $(0, 0, -1)$.

2. Find the critical points of the function f defined by the equation

$$f(u, x, y, z) = ux - x^3u + 2xy^2z - 3u^2yz$$

for every point $(u, x, y, z) \in \mathbf{R}^4$. After supplying the definition of this function to *Scientific Notebook* we point at the system of equations

$$\frac{\partial f(u,x,y,z)}{\partial u} = 0$$

$$\frac{\partial f(u,x,y,z)}{\partial x} = 0$$

$$\frac{\partial f(u,x,y,z)}{\partial y} = 0$$

$$\frac{\partial f(u,x,y,z)}{\partial z} = 0$$

and we click on Solve and Exact. We obtain: Solution is :

$$\{y = y, z = 0, x = 0, u = 0\}, \quad \{y = 0, x = 0, z = z, u = 0\},$$
$$\{y = 0, z = z, x = 1, u = 0\}, \quad \{y = 0, z = z, x = -1, u = 0\}.$$

The critical points of this function are therefore the points (u, x, y, z) of the form $(0, 0, t, 0)$, $(0, 0, 0, t)$, $(0, 1, 0, t)$ or $(0, -1, 0, 1)$ where t is any real number.

3. Given that

$$f(x, y, z) = xy \cos(x + yz)$$

for every point $(x, y, z) \in \mathbf{R}^3$ we can obtain the critical points of the function f as the points $(x, y, z)$of the form $(0, 0, z)$ or $\left(0, \frac{\pi}{2z}, z\right)$ or $\left(0, \frac{\pi}{2}, z\right)$ where z is any real number (with the understanding that $z \neq 0$ wherever the fraction $\frac{1}{z}$ appears).

4. Given that

$$f(x, y, z) = x + xy \cos(x + yz)$$

for every point $(x, y, z) \in \mathbf{R}^3$ we can obtain the critical points of the function f as the points (x, y, z) of the form

$$\left(0, t, \frac{\pi - \arccos \frac{1}{t}}{t}\right)$$

where t is any nonzero real number.

5. Given that

$$f(u, x, y, z) = uxy - x^4 u - 2x^2 y^2 z - 3u^2 yz + x$$

for every point $(u, x, y, z) \in \mathbf{R}^4$, we can obtain the critical points of the function f as the points (u, x, y, z) of the form

$$\left(-\frac{26}{27}\rho^8, \rho, \frac{1}{3}\rho^3, -\frac{13}{9}\rho^4\right)$$

where ρ is a root of the polynomial expression $338Z^{11} + 81$. Since the only real solution of the equation $338Z^{11} + 81 = 0$ is $Z = -\left(\frac{81}{338}\right)^{\frac{1}{11}}$, the function has one critical point[6]

$$\left(-\frac{1}{39}\left(3^{\frac{10}{11}}\right)\left(338^{\frac{3}{11}}\right), -\frac{1}{338}\left(3^{\frac{4}{11}}\right)\left(338^{\frac{10}{11}}\right), -\frac{1}{338}\left(3^{\frac{1}{11}}\right)\left(338^{\frac{8}{11}}\right), -\frac{1}{78}\left(3^{\frac{5}{11}}\right)\left(338^{\frac{7}{11}}\right)\right)$$

5.2.4 Finding Critical Points Numerically

---

6 The next line can be seen completely in the on-screen version of this book.

5.2.4.1 A Polynomial in Three Variables

Find the critical points of the function f defined by the equation

$$f(x, y, z) = x^4 - x^3 y + yz^2 - x + 2xy - 3z$$

for every point $(x, y, z) \in \mathbf{R}^3$. As before, we supply the definition of this function to *Scientific Notebook* and we point at the system of equations

$$\frac{\partial f(x,y,z)}{\partial x} = 0$$

$$\frac{\partial f(x,y,z)}{\partial y} = 0$$

$$\frac{\partial f(x,y,z)}{\partial z} = 0$$

and click on Solve and Exact. We see the solution as

$$\left\{ \begin{array}{c} z = -\frac{904}{303}\rho + \frac{1096}{303}\rho^4 + \frac{1024}{303}\rho^6 + \frac{716}{303}\rho^3 - \frac{256}{101}\rho^8 + \frac{256}{101}\rho^5 - \frac{81}{101} + \frac{147}{202}\rho^2 - \frac{96}{101}\rho^7, \\ y = -\frac{11297}{3232}\rho^3 + \frac{162}{101}\rho^8 + \frac{288}{101}\rho^7 - \frac{216}{101}\rho^6 - \frac{465}{101}\rho^5 - \frac{288}{101}\rho^4 + \frac{587}{202} - \frac{2067}{808}\rho^2 + \frac{3051}{1616}\rho, \\ x = \rho \end{array} \right\}$$

where ρ is a root of

$$64Z^9 - 32Z^6 + 4Z^3 - 17Z^4 + 108Z^2 - 36 - 128Z^7 - 8Z.$$

In order to read this solution you will have to use the horizontal scroll bar at the bottom of your screen. After scrolling to the right, don't forget to scroll back to the left margin. We now separate the formulas for x, y and z into three different lines so that we can work with them more easily:

$x = \rho$

$y = -\frac{11297}{3232}\rho^3 + \frac{162}{101}\rho^8 + \frac{288}{101}\rho^7 - \frac{216}{101}\rho^6 - \frac{465}{101}\rho^5 - \frac{288}{101}\rho^4 + \frac{587}{202} - \frac{2067}{808}\rho^2 + \frac{3051}{1616}\rho$

$z = -\frac{904}{303}\rho + \frac{1096}{303}\rho^4 + \frac{1024}{303}\rho^6 + \frac{716}{303}\rho^3 - \frac{256}{101}\rho^8 + \frac{256}{101}\rho^5 - \frac{81}{101} + \frac{147}{202}\rho^2 - \frac{96}{101}\rho^7$

We point at each of the latter three lines and click on Define and New Definition Then we point at the equation

$$64Z^9 - 32Z^6 + 4Z^3 - 17Z^4 + 108Z^2 - 36 - 128Z^7 - 8Z = 0$$

and click on Solve and Numeric which gives us: Solution is:

$$\{Z = -1.3526\}, \ \{Z = -.55312\}, \ \{Z = 1.4897\}.$$

We are now going to substitute these three numbers for ρ one at a time and, in each case, we evaluate x, y and z to determine a critical point of the function f.

First we point at the equation $\rho = -1.3526$ and click on Define and New Definition. We drag down a copy of the equations

$$x = \rho$$
$$y = -\frac{11297}{3232}\rho^3 + \frac{162}{101}\rho^8 + \frac{288}{101}\rho^7 - \frac{216}{101}\rho^6 - \frac{465}{101}\rho^5 - \frac{288}{101}\rho^4 + \frac{587}{202} - \frac{2067}{808}\rho^2 + \frac{3051}{1616}\rho$$
$$z = -\frac{904}{303}\rho + \frac{1096}{303}\rho^4 + \frac{1024}{303}\rho^6 + \frac{716}{303}\rho^3 - \frac{256}{101}\rho^8 + \frac{256}{101}\rho^5 - \frac{81}{101} + \frac{147}{202}\rho^2 - \frac{96}{101}\rho^7$$

and for each of these three equations we highlight the right side, hold down your control key and click on Evaluate Numerically. This gives us

$$x = -1.3526$$
$$y = -3.1239$$
$$z = -.47959$$

which gives us the critical point $(-1.3526, -3.1239, -.47959)$.
Repeating this procedure for the solution $\rho = -.55312$ we obtain

$$x = -.55312$$
$$y = 1.5496$$
$$z = .96798$$

which gives us the critical point $(-.55312, 1.5496, .96798)$.
Finally, repeating the procedure for the solution $\rho = 1.4897$ we obtain

$$x = 1.4897$$
$$y = 2.621$$
$$z = .57447$$

which gives us the critical point $(1.4897, 2.621, .57447)$.

5.2.4.2 A Transcendental Function

When a function contains exponential, logarithmic or trigonometric expressions, its critical points can often not be found by clicking on Solve and Exact. Instead, we must use Solve and Numeric. One difficulty in this approach is that *Scientific Notebook* will usually give only one solution and we may have to look around a little to see if there are any others. In this example we shall take

$$f(u, x, y, z) = e^x yz - x^2 u + xy + u^2 + 2x - 3z$$

and we supply this definition to *Scientific Notebook*. If we point at the system of equations

$$\frac{\partial f(u,x,y,z)}{\partial u} = 0$$

$$\frac{\partial f(u,x,y,z)}{\partial x} = 0$$

$$\frac{\partial f(u,x,y,z)}{\partial y} = 0$$

$$\frac{\partial f(u,x,y,z)}{\partial z} = 0$$

and click on Solve and Numeric we see: Solution is :

$$\{x = 1.2179, \quad u = .74167, \quad z = -.36032, \quad y = .88753\}$$

and we deduce that the point $(.74167, 1.2179, .88753, -.36032)$ is a critical point of f. Now try pointing at the system

$$\frac{\partial f(u,x,y,z)}{\partial u} = 0$$

$$\frac{\partial f(u,x,y,z)}{\partial x} = 0$$

$$\frac{\partial f(u,x,y,z)}{\partial y} = 0$$

$$\frac{\partial f(u,x,y,z)}{\partial z} = 0$$

$$u \in [-1, 8]$$

$$x \in [-1, 8]$$

$$y \in [-1, 8]$$

$$z \in [-1, 8]$$

and again clicking on Solve and Numeric. You will obtain the same solution as before. You will notice, however, that by replacing the interval $[-1, 8]$ by an interval such as $[-10, 0]$ that does not contain this solution and clicking on Solve and Numeric that you will obtain no solution at all to the system. Try again using the interval $[2, 100]$. Once again you will find that there is no solution. This leads us to believe that the point $(.74167, 1.2179, .88753, -.36032)$ is the only critical point of the function f.

5.2.4.3 Another Transcendental Function

In this example we take

$$f(x, y, z) = x \cos yz + y \sin(x + z) + 2z \sin(x - y)$$

and we supply the definition to *Scientific Notebook* by clicking on Define and New Definition. If we point at the system of equations

$$\frac{\partial f(x,y,z)}{\partial x} = 0$$

$$\frac{\partial f(x,y,z)}{\partial y} = 0$$

$$\frac{\partial f(x,y,z)}{\partial z} = 0$$

and click on Solve and Numeric then we obtain the solution: Solution is :

$$\{x = 2.3037, \quad y = -2.1572, \quad z = 1.4753\}$$

from which we deduce that the point $(2.3037, -2.1572, 1.4753)$ is a critical point of the function f.

Now we point at the system

$$\frac{\partial f(x,y,z)}{\partial x} = 0$$

$$\frac{\partial f(x,y,z)}{\partial y} = 0$$

$$\frac{\partial f(x,y,z)}{\partial z} = 0$$

$$x \in [0, 2]$$

$$y \in [0, 10]$$

$$z \in [1, 2]$$

and click on Solve and Numeric. This time we obtain: Solution is:

$$\{z = 1.4958, \quad y = 5.1596, \quad x = .22806\}$$

from which we deduce that the point $(.22806, 5.1596, 1.4958)$ is another critical point of f. You should try a variety of intervals to see what other critical points you can find.

5.2.5 The Second Derivative Test

We shall state this test for functions of two variables. Suppose that f is a function that has continuous partial derivatives of order two in a region of $\mathbf{R}^2$. For each point (x, y) in this region, we write

$$\Delta(x,y) = \left(\frac{\partial^2 f(x,y)}{\partial x^2}\right)\left(\frac{\partial^2 f(x,y)}{\partial y^2}\right) - \left(\frac{\partial^2 f(x,y)}{\partial x \partial y}\right)^2$$

and

$$T(x,y) = \frac{\partial^2 f(x,y)}{\partial x^2}$$

Suppose that (x, y) is a critical point of the function f.

1. If $\Delta(x,y) > 0$ and $T(x,y) > 0$ then the function f has a local minimum at the point (x, y).
2. If $\Delta(x,y) > 0$ and $T(x,y) < 0$ then the function f has a local maximum at the point (x, y).
3. If $\Delta(x,y) < 0$ then the function f has neither a local maximum nor a local minimum at the point (x, y). A point of this type is known as a **saddle point** of the function f.

5.2.6 An Optimization Example

In this example we take

$$f(x,y) = e^{-x^2-2y^2}\left(x^2 - 3xy + 2y^2 + x - y + 2\right)$$

for each point $(x,y) \in \mathbf{R}^2$ and we begin by pointing at this equation and clicking on Define and New Definition.

5.2.6.1 Preliminary Observations about the Function f

We begin our study of the function f by observing that $f(x,y) \to 0$ as either x or y approaches $\pm\infty$. Next we observe that the expression $f(x,y)$ is never equal to 0. In fact, the condition $f(x,y) = 0$ requires that

$$x^2 - 3xy + 2y^2 + x - y + 2 = 0$$

and, by pointing at this equation and clicking on Solve and Exact and filling x into the dialog box

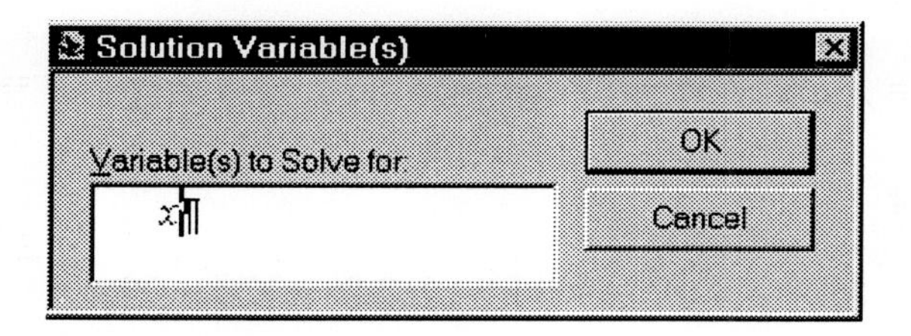

that appears, we obtain the values of x as

$$\left\{x = \frac{3}{2}y - \frac{1}{2} + \frac{1}{2}\sqrt{(y^2 - 2y - 7)}\right\}, \quad \left\{x = \frac{3}{2}y - \frac{1}{2} - \frac{1}{2}\sqrt{(y^2 - 2y - 7)}\right\}.$$

This solution is never a real number because the quadratic $y^2 - 2y - 7$ is always negative.

Since $f(0,0) = 2 > 0$ we can therefore conclude that $f(x,y) > 0$ at every point (x,y) and it follows that the function f has no minimum. On the other hand, by pointing at the expression $f(x,y)$ and clicking on Plot 3D and Rectangular (and adjusting the graph a little) we see the graph of the function f as in the next figure:

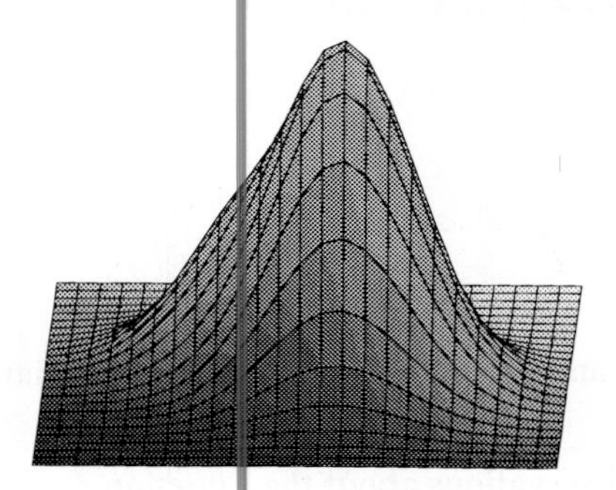

and we can see that the function f does have a maximum somewhere.

5.2.6.2 Finding the Critical Points of f

To find the critical points of f, we point at the system of equations

$$\frac{\partial f(x,y)}{\partial x} = 0$$

$$\frac{\partial f(x,y)}{\partial y} = 0$$

and click on Solve and Exact. We obtain
Solution is :

$$\left\{y = \frac{72}{737}\rho^3 - \frac{27}{5159}\rho^4 - \frac{69}{938}\rho^2 - \frac{7463}{10318}\rho + \frac{225}{5159}, \quad x = \rho\right\}$$

where ρ is a root of

$$36Z^5 - 60Z^4 - 600Z^3 - 180Z^2 + 17Z + 59 = 0$$

Now we point at the latter polynomial equation and click on Solve and Numeric and change the symbol Z to ρ again. We obtain: Solution is :

$$\{\rho = -3.1382\}, \quad \{\rho = .39389\}, \quad \{\rho = 5.1122\}$$

For each of these three values of ρ we want to work out the critical point

$$\left(\rho, \frac{72}{737}\rho^3 - \frac{27}{5159}\rho^4 - \frac{69}{938}\rho^2 - \frac{7463}{10318}\rho + \frac{225}{5159}\right).$$

An efficient way to do these three calculations is to put them in a table. We click on the button [table icon] in the toolbar to make a table, selecting four rows and two columns and fill them in as follows:

ρ	$\left(\rho, \frac{72}{737}\rho^3 - \frac{27}{5159}\rho^4 - \frac{69}{938}\rho^2 - \frac{7463}{10318}\rho + \frac{225}{5159}\right)$
$\rho = -3.1382$	$\left(\rho, \frac{72}{737}\rho^3 - \frac{27}{5159}\rho^4 - \frac{69}{938}\rho^2 - \frac{7463}{10318}\rho + \frac{225}{5159}\right)$
$\rho = .39389$	$\left(\rho, \frac{72}{737}\rho^3 - \frac{27}{5159}\rho^4 - \frac{69}{938}\rho^2 - \frac{7463}{10318}\rho + \frac{225}{5159}\right)$
$\rho = 5.1122$	$\left(\rho, \frac{72}{737}\rho^3 - \frac{27}{5159}\rho^4 - \frac{69}{938}\rho^2 - \frac{7463}{10318}\rho + \frac{225}{5159}\right)$

Now for each of the rows 2, 3 and 4, we point at the equation that gives the value of ρ and click on Define and New Definition. Then we highlight the expression

$$\left(\rho, \frac{72}{737}\rho^3 - \frac{27}{5159}\rho^4 - \frac{69}{938}\rho^2 - \frac{7463}{10318}\rho + \frac{225}{5159}\right),$$

hold down the control key and click on Evaluate. The table becomes

ρ	$\left(\rho, \frac{72}{737}\rho^3 - \frac{27}{5159}\rho^4 - \frac{69}{938}\rho^2 - \frac{7463}{10318}\rho + \frac{225}{5159}\right)$
$\rho = -3.1382$	$(-3.1382, -1.9379)$
$\rho = .39389$	$(.39389, -.24686)$
$\rho = 5.1122$	$(5.1122, 3.9012)$

5.2.6.3 Preparing for the Second Derivative Test

In order to use the second derivative test we need to supply *Scientific Notebook* with the definitions of $\Delta(x,y)$ and $T(x,y)$. It is not good enough to give these definitions directly as

$$\Delta(x,y) = \left(\frac{\partial^2 f(x,y)}{\partial x^2}\right)\left(\frac{\partial^2 f(x,y)}{\partial y^2}\right) - \left(\frac{\partial^2 f(x,y)}{\partial x \partial y}\right)^2$$

and

$$T(x,y) = \frac{\partial^2 f(x,y)}{\partial x^2}$$

because these definitions will not work once we have supplied *Scientific Notebook* with the values of x and y at a particular critical point. We therefore point at each of the equations

$$\Delta(x,y) = \left(\frac{\partial^2 f(x,y)}{\partial x^2}\right)\left(\frac{\partial^2 f(x,y)}{\partial y^2}\right) - \left(\frac{\partial^2 f(x,y)}{\partial x \partial y}\right)^2$$

and

$$T(x,y) = \frac{\partial^2 f(x,y)}{\partial x^2}$$

and, while holding down the control key, we click on Evaluate. This gives us:

$\Delta(x,y) = -\frac{32}{e^{2x^2}e^{4y^2}}x^5 + \frac{4}{e^{x^2}e^{2y^2}}y - \frac{2}{e^{x^2}e^{2y^2}}x - \frac{6}{e^{x^2}e^{2y^2}}x^2 - \frac{12}{e^{x^2}e^{2y^2}}y^2 + \frac{16}{e^{2x^2}e^{4y^2}}x^3 - \frac{24}{e^{2x^2}e^{4y^2}}x^4 + \frac{40}{e^{2x^2}e^{4y^2}}x^2 + \frac{56}{e^{2x^2}e^{4y^2}}y^2 - \frac{32}{e^{2x^2}e^{4y^2}}y^3 - \frac{64}{e^{2x^2}e^{4y^2}}y^4 + \frac{128}{e^{2x^2}e^{4y^2}}y^5 - \frac{128}{e^{2x^2}e^{4y^2}}y^6 + \frac{32}{e^{2x^2}e^{4y^2}}x - \frac{32}{e^{2x^2}e^{4y^2}}y + \frac{3}{e^{x^2}e^{2y^2}} + \frac{8}{e^{2x^2}e^{4y^2}} - \frac{8}{e^{x^2}e^{2y^2}}x^3y + \frac{24}{e^{x^2}e^{2y^2}}x^2y^2 - \frac{8}{e^{x^2}e^{2y^2}}x^2y - \frac{16}{e^{x^2}e^{2y^2}}y^3x + \frac{8}{e^{x^2}e^{2y^2}}y^2x - \frac{32}{e^{2x^2}e^{4y^2}}x^3y + \frac{472}{e^{2x^2}e^{4y^2}}x^2y^2 - \frac{304}{e^{2x^2}e^{4y^2}}x^2y - \frac{464}{e^{2x^2}e^{4y^2}}x^4y^2 + \frac{640}{e^{2x^2}e^{4y^2}}x^3y^3 - \frac{992}{e^{2x^2}e^{4y^2}}x^2y^4 - \frac{480}{e^{2x^2}e^{4y^2}}x^3y^2 + \frac{768}{e^{2x^2}e^{4y^2}}x^2y^3 - \frac{64}{e^{2x^2}e^{4y^2}}xy^3 + \frac{320}{e^{2x^2}e^{4y^2}}xy^2 - \frac{224}{e^{2x^2}e^{4y^2}}xy + \frac{768}{e^{2x^2}e^{4y^2}}xy^5 - \frac{640}{e^{2x^2}e^{4y^2}}xy^4 - \frac{16}{e^{2x^2}e^{4y^2}}x^6 + \frac{192}{e^{2x^2}e^{4y^2}}x^5y + \frac{256}{e^{2x^2}e^{4y^2}}x^4y + \frac{64}{e^{2x^2}e^{4y^2}}x^6y^2 - \frac{384}{e^{2x^2}e^{4y^2}}x^5y^3 + \frac{832}{e^{2x^2}e^{4y^2}}x^4y^4 + \frac{128}{e^{2x^2}e^{4y^2}}x^5y^2 - \frac{512}{e^{2x^2}e^{4y^2}}x^4y^3 - \frac{768}{e^{2x^2}e^{4y^2}}x^3y^5 + \frac{640}{e^{2x^2}e^{4y^2}}x^3y^4 + \frac{256}{e^{2x^2}e^{4y^2}}x^2y^6 - \frac{256}{e^{2x^2}e^{4y^2}}x^2y^{52}$

and[7]

$T(x,y) = -2e^{-x^2-2y^2}\left(x^2 - 3xy + 2y^2 + x - y + 2\right) + 4x^2e^{-x^2-2y^2}\left(x^2 - 3xy + 2y^2 + x - y + 2\right) - 4xe^{-x^2-2y^2}(2x - 3y + 1) + 2e^{-x^2-2y^2}$.

(Actually, we also chose to click on Expand for the expression $\Delta(x,y)$ because the expanded form fits more conveniently onto the computer screen.) Then we point at each of the two latter equations and click on Define and New Definition.

5.2.6.4 Applying the Second Derivative Test

First we point at each of the equations

$$x = \rho$$

$$y = \frac{72}{737}\rho^3 - \frac{27}{5159}\rho^4 - \frac{69}{938}\rho^2 - \frac{7463}{10318}\rho + \frac{225}{5159}x$$

and click on Define and New Definition. Now we click on the button [table button] again to make a table which we fill in as

ρ	$\rho = -3.1382$	$\rho = .39389$	$\rho = 5.1122$
(x,y)	(x,y)	(x,y)	(x,y)
$f(x,y)$	$f(x,y)$	$f(x,y)$	$f(x,y)$
$\Delta(x,y)$	$\Delta(x,y)$	$\Delta(x,y)$	$\Delta(x,y)$
$T(x,y)$	$T(x,y)$	$T(x,y)$	$T(x,y)$
Conclusion			

For each of the columns 2, 3 and 4, we highlight the equation that gives the value of ρ and click on Define and New Definition. Then we point at the expressions (x,y), $f(x,y)$,

7 The next line can be seen completely in the on-screen version of this book.

$\Delta(x,y)$ and $T(x,y)$ in this column, hold down the control key, and click on Evaluate. The table becomes

ρ	$\rho = -3.1382$	$\rho = .39389$	$\rho = 5.1122$
(x,y)	$(-3.1382, -1.9379)$	$(.39389, -.24686)$	$(5.1122, 3.9012)$
$f(x,y)$	-2.4756×10^{-9}	2.4329	-1.262×10^{-26}
$\Delta(x,y)$	4.2773×10^{-14}	43.938	6.4725×10^{-48}
$T(x,y)$	1.6022×10^{-7}	-4.8596	1.8852×10^{-24}
Conclusion	Local minimum?	Local maximum	Local minimum?

From this table we see that there is a local maximum only when $\rho = .39389$ and we conclude that the function f has a maximum value at the critical point $(.39389, -.24686)$. At each of the other two critical points the function f has a local minimum if we can trust the numbers in the preceding table. However, these numbers ought to be treated with suspicion because they are so close to zero. Since the value of the function f is smaller at each of these two points we can say with certainty that f has its maximum value at the point $(.39389, -.24686)$.

5.2.7 Another Optimization Example

In this example we take

$$f(x,y) = x^3 - 3xy + y^3$$

for every point (x,y) that lies on or inside the ellipse

$$\frac{x^2}{4} + \frac{y^2}{9} = 1.$$

We point to this definition of $f(x,y)$ and click on Define and New Definition.

The graph of this function can be pictured as in the next figure:

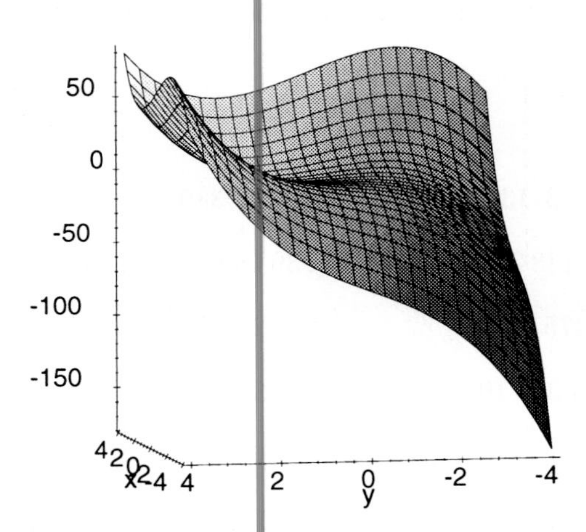

Since the domain of this continuous function is a closed bounded set in $\mathbf{R}^2$ we know that f has both a maximum and a minimum value. From Fermat's theorem we know that a local extremum of this function f must either occur at a critical point inside the ellipse $\frac{x^2}{4}+\frac{y^2}{9}=1$ or must occur at a point on the ellipse itself.

5.2.7.1 Finding the Critical Points of f

To find the critical points of f, we point at the system of equations

$$\begin{array}{l}\frac{\partial f(x,y)}{\partial x}=0\\ \frac{\partial f(x,y)}{\partial y}=0\end{array}$$

and click on Solve and Exact. We obtain the solution as

$$\{x=0,y=0\},\ \{y=1,x=1\},\ \{x=-\rho-1,y=\rho\}$$

where ρ is a root of Z^2+Z+1. Since the polynomial Z^2+Z+1 has no real roots we conclude that the only critical points of the function f are the points $(0,0)$ and $(1,1)$.

5.2.7.2 Preparing for the Second Derivative Test

In order to use the second derivative test we need to supply *Scientific Notebook* with the definitions of $\Delta(x,y)$ and $T(x,y)$. It is not good enough to give these definition directly as

$$\Delta(x,y)=\left(\frac{\partial^2 f(x,y)}{\partial x^2}\right)\left(\frac{\partial^2 f(x,y)}{\partial y^2}\right)-\left(\frac{\partial^2 f(x,y)}{\partial x\partial y}\right)^2$$

and

$$T(x,y)=\frac{\partial^2 f(x,y)}{\partial x^2}$$

because these definitions will not work once we have supplied *Scientific Notebook* with the values of x and y at a particular critical point. We therefore point at each of the equations

$$\Delta(x, y) = \left(\frac{\partial^2 f(x, y)}{\partial x^2}\right)\left(\frac{\partial^2 f(x, y)}{\partial y^2}\right) - \left(\frac{\partial^2 f(x, y)}{\partial x \partial y}\right)^2$$

and

$$T(x, y) = \frac{\partial^2 f(x, y)}{\partial x^2}$$

and, while holding down the control key, we click on Evaluate. This gives us:

$$\Delta(x, y) = 36xy - 9$$

and

$$T(x, y) = 6x.$$

Then we point at each of the two latter equations and click on Define and New Definition.

5.2.7.3 Applying the Second Derivative Test

We click on the button [table button] in the toolbar to make a table which we fill in as

x	y	$\Delta(x, y)$	$T(x, y)$	conclusion
$x = 0$	$y = 0$	$\Delta(x, y)$	$T(x, y)$	
$x = 1$	$y = 1$	$\Delta(x, y)$	$T(x, y)$	

For each of the rows 2 and 3, we point at the equations that give the values of x and y and click on Define and New Definition. Then we point at the expressions $\Delta(x, y)$ and $T(x, y)$ in this row, hold down the control key, and click on Evaluate. The table becomes

x	y	$\Delta(x, y)$	$T(x, y)$	conclusion
$x = 0$	$y = 0$	-9	0	saddle point
$x = 1$	$y = 1$	27	6	local minimum

We conclude that the function f has neither a maximum nor a minimum value at the point $(0, 0)$ and that it may possibly have its minimum value at $(1, 1)$. Note that $f(0, 0) = 0$ and $f(1, 1) = -1$.

5.2.7.4 Behavior of f on the Boundary

We must now examine the behavior of the function f on the ellipse

$$\frac{x^2}{4} + \frac{y^2}{9} = 1.$$

The simplest way to describe this ellipse is to describe it parametrically as

$$\{(2\cos\theta, 3\sin\theta) \mid \theta \in [0, 2\pi]\}.$$

With this in mind we define

$$g(\theta) = 8\cos^3\theta + 18\cos\theta\sin\theta + 27\sin^3\theta$$

for $0 \le \theta \le 2\pi$ and we point at the definition of $g(\theta)$ and click on Define and New Definition.

$g'(\theta) = 0$, Solution is

$$\{\theta = 2\arctan(\rho)\}$$

where ρ is a root of

$$8Z - 16Z^3 + 8Z^5 - 69Z^2 + 39Z^4 + 3 + 3Z^6 = 0,$$

Solution is :

$$\{Z = -1.2583\}, \ \{Z = -.16155\}, \ \{Z = .27118\}, \ \{Z = 1.2257\}.$$

In order to examine the values of g at these critical numbers and at the endpoints of the interval $[0, 2\pi]$ we construct the column matrix

$$\begin{matrix} 0 \\ -1.2583 \\ -.16155 \\ .27118 \\ 1.2257 \\ 2\pi \end{matrix}$$

and we evaluate the function g at this matrix by pointing at the expression

$$g\begin{pmatrix} 0 \\ -1.2583 \\ -.16155 \\ .27118 \\ 1.2257 \\ 2\pi \end{pmatrix}$$

and clicking on Evaluate Numerically. This yields

$$g\begin{pmatrix} 0 \\ -1.2583 \\ -.16155 \\ .27118 \\ 1.2257 \\ 2\pi \end{pmatrix} = \begin{pmatrix} 8.0 \\ -17.766 \\ 10.437 \\ 3.0281 \\ 17.08 \\ 8.0 \end{pmatrix}$$

from which we can deduce that both the maximum and minimum values of the function f are achieved on the boundary of the domain. The maximum value of f is 17.08 and the minimum is -17.766.

5.2.8 An Exercise on Maxima and Minima

Given that

$$f(x,y) = \frac{x^2y + 2xy - y^2}{(1 + x^2 + y^2)^2}$$

whenever $-3 \le x \le 3$ and $-3 \le y \le 3$, obtain a sketch of the graph of f, find its nine critical points (some of them numerically) and, at each critical point, determine whether the function f has a local maximum or a local minimum at that point. Find the maximum and minimum values of the function f.

5.2.9 Maximizing a Probability Function

In this example, we suppose that we have n white balls and n black balls which we are going to place in two urns A and B in any way we please, as long as at least one ball is placed into each urn. After this has been done, a second person walks into the room and selects one ball at random. Our problem is to maximize the probability that this person draws a white ball.

We suppose that the distribution of the balls in the urns A and B is as described in the following table:

	A	B
Number of White Balls	x	$n-x$
Number of Black Balls	y	$n-y$

If $P(x,y,n)$ is the probability that a single ball drawn at random will be white then

$$P(x,y,n) = \frac{1}{2}\left(\frac{x}{x+y} + \frac{n-x}{2n-x-y}\right).$$

From now on we shall assume that $n = 50$. We begin our study of the function by looking at the following table which shows the values of $P(x, y, 50)$ at a few selected points (x, y).

$$
\begin{array}{ll}
P(0,1,50) = .25253 & P(1,0,50) = .74747 \\
P(1,1,50) = .5 & P(2,1,50) = .58076 \\
P(1,2,50) = .41924 & P(25,25,50) = .5 \\
P(50,1,50) = .4902 & P(1,50,50) = .5098 \\
P(50,0,50) = .4902 & P(50,49,50) = .25253 \\
P(49,50,50) = .74747 & P(49,49,50) = .5
\end{array}
$$

To solve the problem we need to find the maximum value of the expression $P(x, y, 50)$ as the point (x, y) varies through the rectangle $[0, 50] \times [0, 50]$ from which the points $(0, 0)$ and $(50, 50)$ have been removed. If we sketch the graph $z = P(x, y, 50)$ then we obtain the following surface:

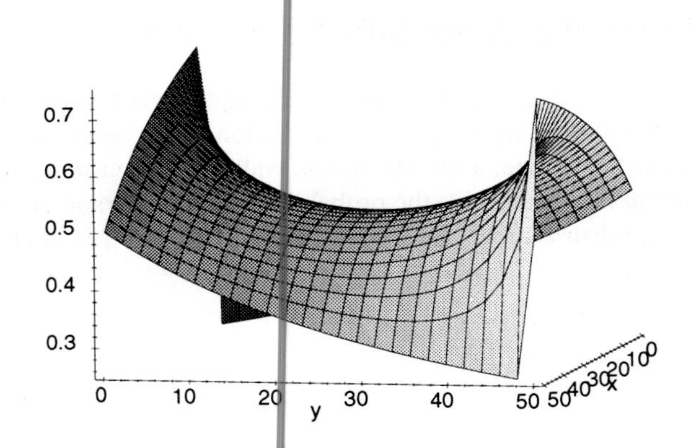

From the looks of this surface it seems unlikely that the maximum value of z will be achieved at a critical point. The maximum appears to be at the left or right extremities of the figure. As a matter of fact, if we point at the equations

$$
\begin{aligned}
\frac{\partial}{\partial x} P(x, y, 50) &= 0 \\
\frac{\partial}{\partial y} P(x, y, 50) &= 0
\end{aligned}
$$

and click on Solve and Exact then we obtain

$$\{y = 25, x = 25\}.$$

As we have already seen, the maximum value of z does not occur at the point $(25, 25)$.

We now examine the boundary behavior of the function. There are four cases to consider

5.2.9.1 The Case $x = 0$ and $1 \le y \le 50$

We define $g(y) = P(0, y, 50)$ for $1 \le y \le 50$. Point at this definition of $g(y)$ and click on Define and New Definition. Since

$$g(y) = \frac{25}{100 - y}$$

for each y we see that the maximum value of $g(y)$ is $g(50) = \frac{1}{2}$.

5.2.9.2 The Case $y = 0$ and $1 \le x \le 50$

We define $g(x) = P(x, 0, 50)$ for $1 \le x \le 50$. Point at this definition of $g(y)$ and click on Define and New Definition. Since

$$g(x) = \frac{1}{2} + \frac{1}{2}\frac{50 - x}{100 - x} = 1 - \frac{25}{100 - x}$$

for each x, we see that the maximum value of this function is $g(1) = .74747$.

5.2.9.3 The Case $x = 50$ and $0 \le y \le 49$

We define $g(y) = P(50, y, 50)$ for $0 \le y \le 49$. Point at this definition of $g(y)$ and click on Define and New Definition. Since

$$g(y) = \frac{25}{50 + y}$$

for each y, we see that the maximum value of this function is $g(0) = .5$.

5.2.9.4 The Case $y = 50$ and $0 \le x \le 49$

We define $g(x) = P(x, 50, 50)$ for $0 \le x \le 49$. Point at this definition of $g(y)$ and click on Define and New Definition. Since

$$g(x) = 1 - \frac{50}{x + 50}$$

for each x, we see that the maximum value of this function is $g(49) = .74747$.

5.2.9.5 Conclusion

We conclude that the expression $P(x, y, 50)$ takes a maximum value of .74747 at the point $(0, 1)$ and again at the point $(49, 50)$. This means that we can maximize the probability that a white ball will be selected by placing no white ball and just one black ball in urn A and all the other balls in urn B. Alternatively we can place no white ball and just one black ball in urn B and all the other balls in urn A.

5.2.9.6 Some Variations of the Probability Problem

Some of the variations suggested here may be suitable for presentation in the classroom.

Others may be suitable for student projects.

1. Repeat the preceding probability problem assuming that the selection of the ball will be made in such a way that the probability that the selection will be made from urn A is $\frac{1}{3}$ and the probability that the selection will be made from urn B is $\frac{2}{3}$.
2. Extend the preceding variation to the general case in which the probability of selecting a ball from urn A is some number γ satisfying $0 < \gamma < 1$ and the probability of selecting the ball from urn B is $1 - \gamma$.
3. Investigate the problem of determining how the balls should be placed in order to minimize the probability that the selected ball be white. Of course, this is simply the problem of maximizing the probability that a black ball be selected.
4. Suppose that the selection of the ball results in payoffs as described in the following table

	From urn A	From urn B
white ball	α_A	α_B
black ball	β_A	β_B

Study the payoffs that result from different placement of the balls.

5. Determine the maximum value of the expression $P(x, k, n)$ where k is a given integer satisfying $1 \leq k \leq n - 1$. Find the value x_k of x at which this maximum occurs. You will find that

$$x_k = \frac{\sqrt{k}(2n - k) - k\sqrt{n - k}}{\sqrt{n - k} + \sqrt{k}}$$

and that the maximum value of $P(x, k, n)$ is

$$P(x_k, k, n) = \frac{3n - 2\sqrt{(k(n - k))}}{4n}$$

5.2.9.7 Notes on the Probability Problem

Two analytical solutions of this problem are given in D.L. Albig & J.J. Corbet, *Three Especially Interesting Problems for Undergraduates*, Problem Corner, Virginia Mathematics Teacher, Spring 1988. One of these solutions relies on precalculus techniques while the other relies on multivariate calculus. It may be instructive to investigate these two methods with the help of *Scientific Notebook*.

In a talk given by J.J. Corbet and D.L. Albig at the fall 1986 meeting of MD-DC-VA Section of the MAA it was shown algebraically that

$$P(1, 0, n) = \frac{3n - 2}{4n - 2}$$

is the maximum value of $P(x, y, n)$ in the general case. However, the approach taken here illustrates the principle that powerful mathematics computing products such as the operation of Maple in *Scientific Notebook* may be used to suggest some very intuitive and creative approaches to such problems.

The authors would like to thank Stephen P. Corwin for his invaluable suggestions to this section.

5.2.10 Monkey Saddles

The simplest way for a function f to have a saddle point at a critical point (x_1, y_1) is that one of the second partial derivatives $\frac{\partial^2 f}{\partial x^2}$ and $\frac{\partial^2 f}{\partial y^2}$ is positive at A and the other is negative. Suppose, for example that

$$\frac{\partial^2 f(x_1, y_1)}{\partial x^2} > 0 \quad \text{and} \quad \frac{\partial^2 f(x_1, y_1)}{\partial y^2} < 0.$$

Then the function that sends x to $f(x, y_1)$ has a minimum at x_1 and the function that sends y to $f(x_1, y)$ has a maximum at y_1 and we see at once that the function f cannot have a local maximum or minimum at the point (x_1, y_1). A simple example of a function of this type is the function f defined by

$$f(x, y) = x^2 - y^2$$

for all $(x, y) \in \mathbf{R}^2$. Note that $(0, 0)$ is the only critical point of f and that

$$\frac{\partial^2 f(0, 0)}{\partial x^2} = 2$$

and

$$\frac{\partial^2 f(0, 0)}{\partial y^2} = -2.$$

The graph of this function is illustrated in the next figure:

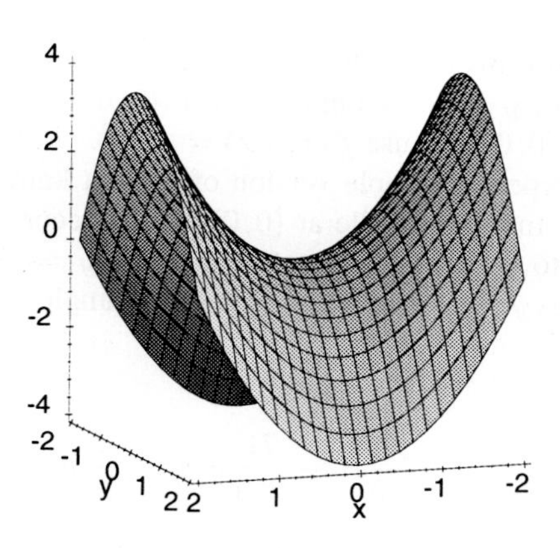

It can quite easily happen, however that a given function f has a saddle point at a given critical point when the second partial derivatives $\frac{\partial^2 f}{\partial x^2}$ and $\frac{\partial^2 f}{\partial y^2}$ have the same sign at this point. The next figure shows the graph of the function f where $f(x,y) = x^2 + y^2 + 4xy$ for all points $(x,y) \in \mathbf{R}^2$. We see at once from Theorem 5.2.5 that, since

$$\left(\frac{\partial^2 f(0,0)}{\partial x^2}\right)\left(\frac{\partial^2 f(0,0)}{\partial y^2}\right) - \left(\frac{\partial^2 f(0,0)}{\partial x \partial y}\right)^2 = (2)(2) - 4^2 < 0,$$

the function f must have a saddle point at $(0,0)$.

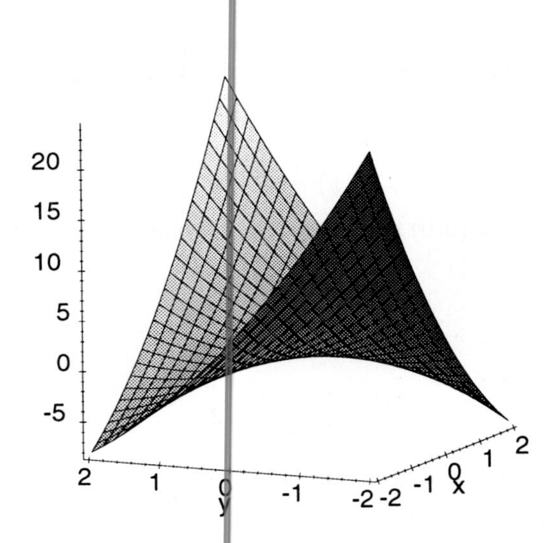

Even though the restriction of this function f to the line $x = 0$ has a minimum at $(0,0)$ and the restriction of f to the line $y = 0$ has a minimum at $(0,0)$, the restriction of f to the line $y = -x$ has a maximum at $(0,0)$ because $f(x,-x) = -2x^2$.

A saddle point of this type is a simple version of a point known as a **monkey saddle.** A more interesting type of monkey saddle at $(0,0)$ would occur when the function has a minimum when restricted to each of the lines $x = 0$ and $y = 0$ and a maximum when restricted to each of the lines $y = x$ and $y = -x$. For an example of a monkey saddle of this type we define

$$f(x,y) = \frac{1}{10}x^4 - \frac{71}{250}x^2y^2 + \frac{1}{10}y^4.$$

The graph of this function is shown in the next figure:

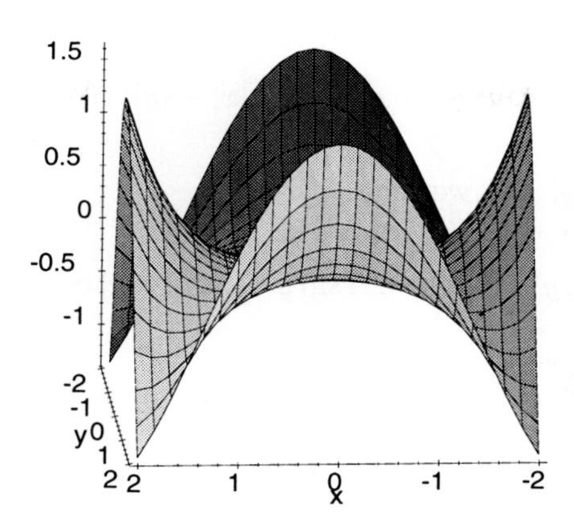

For an animated illustration of this surface as it is rotated about the z-axis, **click here**.

5.2.11 Some Exercises

Analyze each of the following functions for maxima and minima and draw several sketches of its graph taken from a variety of different angles to illustrate the surface.

1. Whenever $|x| \le 4$ and $|y| \le 4$, we define

$$f(x, y) = (1/3)x^3 - 3xy + y^3.$$

2. Whenever $|x| \le 4$ and $|y| \le 4$, we define

$$f(x, y) = \frac{\sin(x + y)}{1 + x^2 + y^2}.$$

3. Whenever $|x| \le 4$ and $|y| \le 4$, we define

$$f(x, y) = \frac{\exp(\sin(x + y))}{1 + x^2 + y^2}.$$

4. Whenever $|x| \le 4$ and $|y| \le 4$, we define

$$f(x, y) = \frac{\sin(x + y)}{\log(2 + x^2 + y^2)}.$$

5. Whenever $|x| \le 4$ and $|y| \le 4$, we define

$$f(x, y) = xy \sin x.$$

6. Whenever $x^2 + y^2 \le 16$, we define

$$f(x, y) = \frac{x^2 + y^2}{3 + \sin x + \sin y}.$$

7. Whenever $|x| \leq 2$ and $|y| \leq 2$, we define

$$f(x, y) = xy \exp(-|x| - |y|).$$

8. Whenever $|x| \leq 3$ and $|y| \leq 3$, we define

$$f(x, y) = (x^2 + y^2) \sin y + (x^2 - y^2) \cos x.$$

9. Whenever $|x| \leq 4$ and $|y| \leq 4$, we define

$$f(x, y) = x^3 - 3xy^2.$$

10. Whenever $|x| \leq 4$ and $|y| \leq 4$, we define

$$f(x, y) = xy(x^2 - y^2).$$

11. Whenever $|x| \leq 1$ and $0 \leq y \leq 2$, we define

$$f(x, y) = (y - x^2)(y - 2x^2).$$

for $-1 \leq x \leq 1$ and $0 \leq y \leq 2$.

12. Whenever $-1 \leq x \leq 3$ and $1 \leq y \leq 4$, we define

$$f(x, y) = -x^4 + 6x^2y - y^2 - x^2y^2.$$

13. Whenever $-1 \leq x \leq .4$ and $1 \leq y \leq .4$, we define

$$z = x^6 + y^6 - xy.$$

14. Draw a family of surfaces of the type

$$f(x, y) = \frac{1}{10}x^4 - cx^2y^2 + \frac{1}{10}y^4$$

for a variety of choices of the constant c and try to determine from your sketches how small c must be in order for the function f to have a minimum at $(0, 0)$. Check your conclusion by applying the second derivative test.

15. $f(x, y) = x^3 - x^2y + y^3$ for $-4 \leq x \leq 4$ and $-4 \leq y \leq 4$. Obtain a sketch of this graph that resembles the following figure:

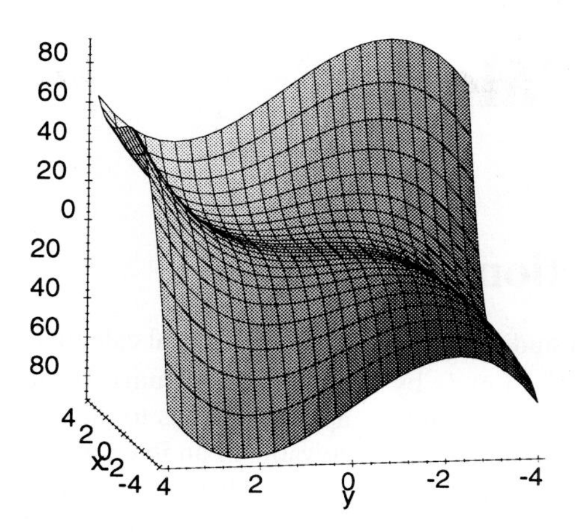

16. Sketch a variety of graphs of the function

$$f(x,y) = \frac{x^3 - y^3}{x^2 + y^2}$$

including one that resembles the following figure:

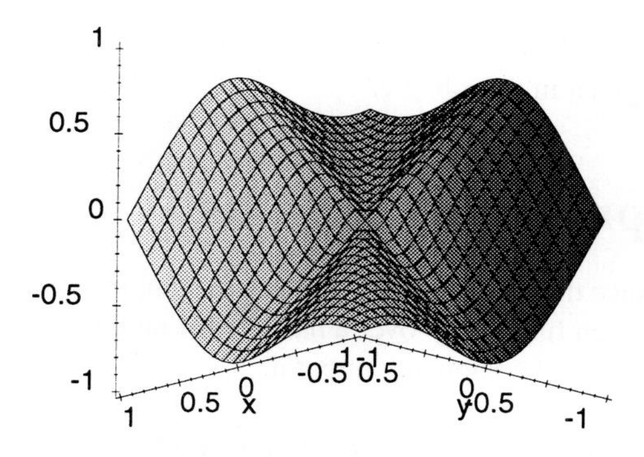

and conclude from your sketches that the function f is not differentiable at $(0, 0)$.

Chapter 6
Integral Calculus

6.1 Introduction

One of the topics that are studied in a first course in integral calculus is the process of approximating a given integral $\int_a^b f(x)\,dx$ by various types of sums. In studying this topic, we are not just concerned with the act of finding approximations to the integral. As we saw in Exercise 4 of Subsection 2.1.4, an approximate value of an integral can be obtained by a single click on Evaluate Numerically. The purpose of this topic is to acquaint students with a variety of different kinds of sums such as **left sums**, **right sums**, **trapezoidal sums**, **midpoint sums** and **Simpson sums,** to point out that some of these sums will approximate a given integral more closely than others and to show that all of them provide better approximations when the interval of integration is more finely partitioned. We would like to know how much better the better sums are and how much better the sums become when the interval is more finely partitioned.

In this chapter we shall show how *Scientific Notebook* can be used to study the left sums, right sums, trapezoidal sums, midpoint sums and Simpson sums of a given function f on an interval $[a,b]$ and how a course in integral calculus can thus be enriched with the help of *Scientific Notebook*.

The on-screen version of this book also provides a link to a more detailed presentation of this topic than we have given in this chapter.

6.2 The Approximating Sums

In this section we introduce the notions of left sum, right sum, trapezoidal sum, midpoint sum and Simpson sum of a given function f over a partition of an interval $[a,b]$. We begin with a brief review of the definition of a Riemann integral.

6.2.1 Brief Review of the Riemann Integral

In a first course in integral calculus, the **Riemann integral** of a bounded function f on an interval $[a,b]$ is described as the limit of a sequence of sums of the type

$$\sum_{j=1}^{n} (x_j - x_{j-1})\, f(t_j) \tag{6.1}$$

where

$$a = x_0 < x_1 < x_2 < \cdots < x_{n-1} < x_n = b$$

and where, for each $j = 1, 2, \cdots, n$ we have

$$x_{j-1} \le t_j \le x_j.$$

Sums of this type are called **Riemann sums** of the function f over the interval $[a, b]$. The sense in which the limit is taken is that if we define the **mesh** of the partition

$$(x_0, x_1, x_2, \cdots, x_n)$$

to be the largest of the lengths of the intervals $[x_{j-1}, x_j]$ then the Sum (6.1) can be made as close as we like to $\int_a^b f(x)\,dx$ by making this mesh small enough.

The simplest type of partition of a given interval $[a, b]$ is a partition $(x_0, x_1, x_2, \cdots, x_n)$ for which all of the intervals $[x_{j-1}, x_j]$ have the same length. In this case the partition is said to be **regular** and for each $j = 1, 2, \cdots, n$ we have

$$x_j - x_{j-1} = \frac{b-a}{n}$$

and

$$x_j = a + \frac{j(b-a)}{n}.$$

Since the mesh of this partition is $\frac{b-a}{n}$, we make it approach 0 by letting $n \to \infty$.

6.2.2 Defining a Partition in *Scientific WorkPlace*

Given $a < b$ and a positive integer n, we have described the regular partition $(x_0, x_1, x_2, \cdots, x_n)$ as being the finite sequence of numbers defined by

$$x_j = a + \frac{j(b-a)}{n}$$

for each $j = 0, 1, \cdots, n$. Before we give this definition to *Scientific Notebook*, we shall make the notation a bit more precise. We shall replace the notation x_j by $x(a, b, j, n)$ in order to account for the fact that this number depends also upon the value of n and upon the interval that is being partitioned. Accordingly, the first step in the procedure is to point at the equation

$$x(a, b, j, n) = a + \frac{j(b-a)}{n}$$

and to click on Define and New Definition.

6.2.3 Introducing a Temporary Function *f*

Since the various approximating sums all depend upon the function f that we wish to integrate, we need to let *Scientific Notebook* know that the symbol f stands for a function *before* we write down the definitions of the approximating sums. In order to achieve this, we make the nominal definition $f(x) = x^2$ by pointing at the equation

$$f(x) = x^2$$

and clicking on Define and New Definition.

Note that this definition of f is purely temporary. We can change it at any time and all the sums will change accordingly.

6.2.4 Defining the Approximating Sums

The **left sum** of a given function f over the partition defined above is the Riemann sum

$$\sum_{j=1}^{n} \left(\frac{b-a}{n}\right) f(t_j)$$

where, for each j, the number t_j is the left endpoint of the interval that runs from $x(a,b,j-1,n)$ to $x(a,b,j,n)$.

$x(a,b,j-1,n)$ $x(a,b,j,n)$

In other words, we define the left sum by pointing at the equation

$$L_f(a,b,n) = \sum_{j=1}^{n} \left(\frac{b-a}{n}\right) f(x(a,b,j-1,n))$$

and clicking on Define and New Definition. Similarly, the right sum of f is

$$R_f(a,b,n) = \sum_{j=1}^{n} \left(\frac{b-a}{n}\right) f(x(a,b,j,n))$$

and we define it by pointing at the equation and clicking on Define and New Definition. The arithmetic mean of the left and right sums is the **trapezoidal sum** $T_f(a,b,n)$ which we define by pointing at the equation

$$T_f(a,b,n) = \frac{1}{2}\left(L_f(a,b,n) + R_f(a,b,n)\right).$$

Alternatively we could observe that

$$T_f(a,b,n) = \frac{b-a}{2n}\left(f(x(a,b,0,n)) + 2\sum_{j=1}^{n-1} f(x(a,b,j,n)) + f(x(a,b,n,n))\right)$$

and use this equation for the definition of the trapezoidal sum. As we shall see from the examples that follow, the trapezoidal sum is frequently a much better approximation to the integral than either the left or the right sum. An even better approximation than the trapezoidal sum is the **midpoint sum** $M_f(a,b,n)$ which we define by pointing at the equation

$$M_f(a,b,n) = \sum_{j=1}^{n} \left(\frac{b-a}{n}\right) f\left(\frac{x(a,b,j-1,n) + x(a,b,j,n)}{2}\right).$$

In this sum the function f is evaluated for each j at the midpoint of the interval that runs from $x(a,b,j-1,n)$ to $x(a,b,j,n)$.

Finally, the **Simpson sum** $S_f(a,b,n)$ of f over the given partition is defined by pointing at the equation[8]

$$S_f(a,b,n) = \frac{b-a}{3n}\left(f(x(a,b,0,n)) + \sum_{j=1}^{n-1}\left(3-(-1)^j\right) f(x(a,b,j,n)) + f(x(a,b,n,n))\right)$$

As you may know, the Simpson sum is used only when the number n is even.

6.3 A Simple Example

Having supplied the definitions of the sums to *Scientific Notebook* as described in the previous section, we can evaluate the sums for any specified function f, interval $[a,b]$ and any specified value of n. In this section we work out some approximations to the integral $\int_1^5 (x^3 - 2x^2 + x - 3)\,dx$. We know, of course, that

$$\int_1^5 (x^3 - 2x^2 + x - 3)\,dx = \frac{220}{3} \approx 73.33333.$$

To work out the various approximating sums, we begin by pointing at the equation $f(x) = x^3 - 2x^2 + x - 3$ and clicking on Define and New Definition.

6.3.1 Approximations with 20 Subdivisions

By pointing and clicking on Evaluate Numerically we obtain

$$\begin{aligned} L_f(1,5,20) &= 65.52 \\ R_f(1,5,20) &= 81.52 \\ T_f(1,5,20) &= 73.52 \\ M_f(1,5,20) &= 73.24 \\ S_f(1,5,20) &= 73.33333 \end{aligned}$$

[8] The next line can be seen completely in the on-screen version of this book.

and we can see at once that the midpoint sum is better than the trapezoidal sum which, in turn, is much better than the left and right sums. We see also that the Simpson sum is the best of all. As a matter of fact, the exact value of the Simpson sum is

$$S_f(1, 5, 20) = \frac{220}{3}$$

which is exactly equal to the integral. It can be proved that the Simpson sum is always exactly correct when the function being integrated is a polynomial of degree 3 or less.

6.3.2 Illustrating the Approximating Sums

Using *Scientific Notebook*, we can easily picture the sums $L_f(1, 5, 20)$, $R_f(1, 5, 20)$, $T_f(1, 5, 20)$ and $M_f(1, 5, 20)$. To obtain these illustrations we point at the name f of the function and click on Calculus and Plot Approx. Integral. After first sketch comes up we open its dialog box and move to Plot Properties. We see the plot properties dialog box

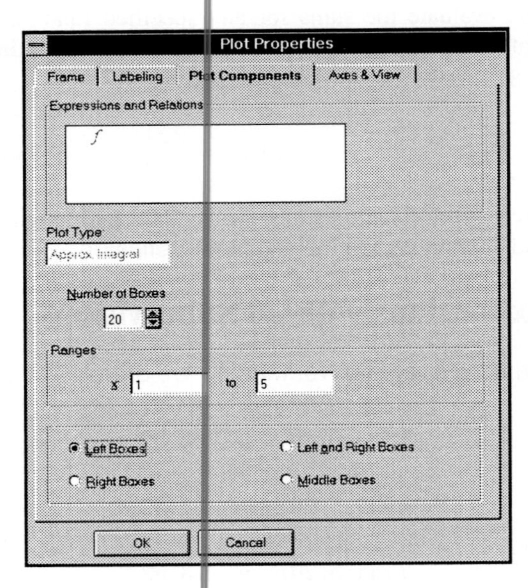

and we fill in 20 as the number of boxes and the range of x as being from 1 to 5. Depending upon whether we have selected left boxes, right boxes (or both), or middle boxes, the figure will illustrate the left sum, right sum or midpoint sum respectively. For example, if both left and right boxes are selected, the plot is shown as in the next figure. We can use this figure to make the point that the trapezoidal sum would be correct if the graph of f were a straight line in each of the 20 columns.

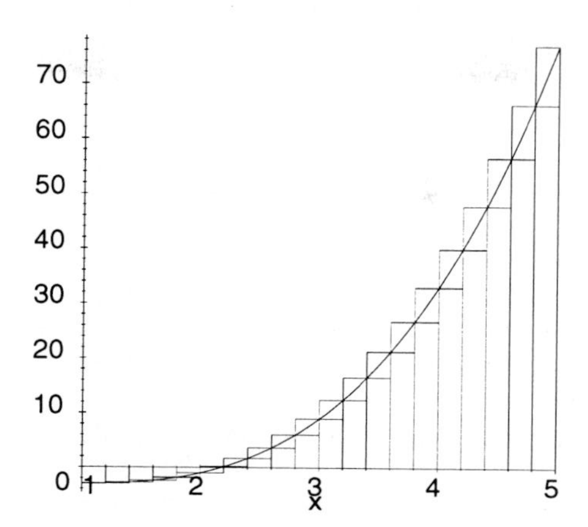

6.3.3 Approximating Sums for a Variety of Values of *n*

An efficient way to obtain approximating sums for a variety of values of n is to type a table like the one that follows:

	$n = 50$	$n = 100$	$n = 250$	$n = 500$
$L_f(a, b, n)$	$L_f(a, b, n)$	$L_f(a, b, n)$	$L_f(a, b, n)$	$L_f(a, b, n)$
$R_f(a, b, n)$	$R_f(a, b, n)$	$R_f(a, b, n)$	$R_f(a, b, n)$	$R_f(a, b, n)$
$T_f(a, b, n)$	$T_f(a, b, n)$	$T_f(a, b, n)$	$T_f(a, b, n)$	$T_f(a, b, n)$
$M_f(a, b, n)$	$M_f(a, b, n)$	$M_f(a, b, n)$	$M_f(a, b, n)$	$M_f(a, b, n)$
$S_f(a, b, n)$	$S_f(a, b, n)$	$S_f(a, b, n)$	$S_f(a, b, n)$	$S_f(a, b, n)$

In order to evaluate the entries in this table we proceed as follows

1. Point at each of the equations $a = 1$ and $b = 5$ and click on Define and New Definition.
2. To obtain the values of the sums in the $n = 50$ column, point at the equation $n = 50$ in the top box of this column and click on Define and New Definition. Then move down the column highlighting the expression that in each box and, while holding down the control button, click on Evaluate Numerically.
3. Repeat this procedure for each of the remaining columns.

The table now appears as

	$n = 50$	$n = 100$	$n = 250$	$n = 500$
$L_f(a,b,n)$	70.1632	71.7408	72.69453	73.01363
$R_f(a,b,n)$	76.5632	74.9408	73.97453	73.65363
$T_f(a,b,n)$	73.3632	73.3408	73.33453	73.33363
$M_f(a,b,n)$	73.3184	73.3296	73.33274	73.33318
$S_f(a,b,n)$	73.33333	73.33333	73.33333	73.33333

Looking at this table we can see that all of the sums improve as n is increased and that the Simpson sum remains the most accurate followed by the midpoint sum, the trapezoidal sum and finally the left and right sums.

6.4 A More Complicated Example

In this section we consider the integral

$$\int_0^1 \sqrt[3]{1 - x^2}\,dx.$$

What makes this integral more difficult to approximate is the fact that the graph $y = \sqrt[3]{1 - x^2}$ is vertical when $x = 1$. See the next figure:

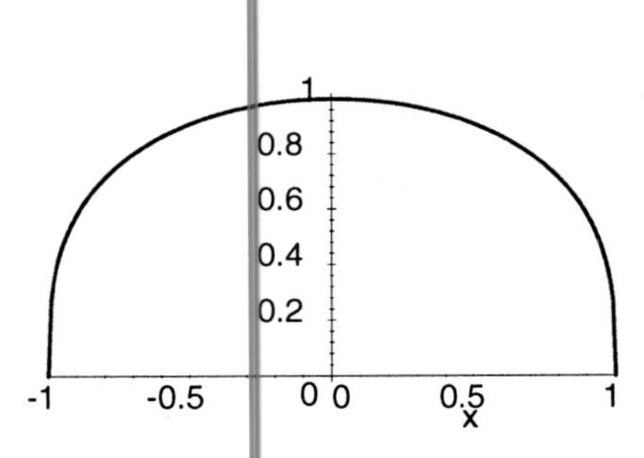

As we can see by clicking on Evaluate Numerically, the value of this integral to seven decimal places is

$$\int_0^1 \sqrt[3]{1 - x^2}\,dx = .8413093.$$

6.4.1 Approximating Sums for a Variety of Values of n

By giving *Scientific Notebook* the definitions $f(x) = \sqrt[3]{1-x^2}$, $a = 0$ and $b = 1$, we can obtain the following table

	$n = 50$	$n = 100$	$n = 200$	$n = 400$
$L_f(a,b,n)$	.8494132	.8455566	.8435105	.8424407
$R_f(a,b,n)$	.8294132	.8355566	.8385105	.8399407
$T_f(a,b,n)$	.8394132	.8405566	.8410105	.8411907
$M_f(a,b,n)$	.8417001	.8414645	.8413709	.8413337
$S_f(a,b,n)$	.840373	.8409378	.8411618	.8412508

Notice how the midpoint sum does a consistently better job than the Simpson sum in this example.

6.4.2 Obtaining Arrays of Approximating Sums

By producing an array of midpoint and Simpson sums we can get a better idea of how many subdivisions of the interval are needed to make these sums approximate the integral to a certain level of accuracy. We begin by defining

$$m(n) = M_f(0,1,n)$$

and

$$s(n) = S_f(0,1,n).$$

Now we want to make a column of numbers and, for this purpose, we define

$$h(x) = x + 100,$$

we click on Calculus and Iterate and we fill in the iterate dialog box as shown.

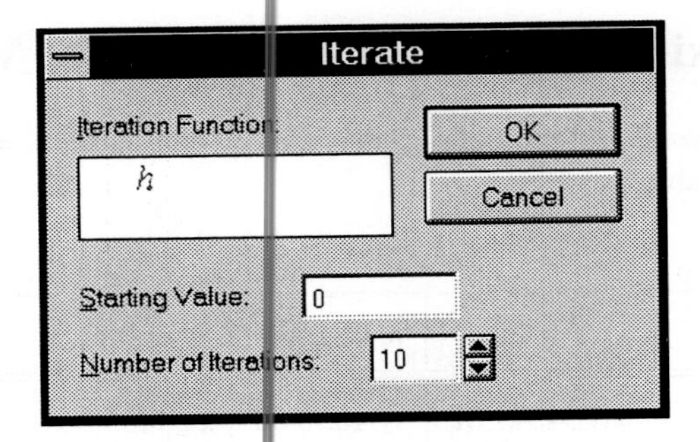

This gives us the following list of iterates of the function h:

$$\begin{pmatrix} 100 \\ 200 \\ 300 \\ 400 \\ 500 \\ 600 \\ 700 \\ 800 \\ 900 \\ 1000 \\ 1100 \end{pmatrix}$$

If we evaluate the functions m and s at this column of numbers, we obtain

$$m\begin{pmatrix} 100 \\ 200 \\ 300 \\ 400 \\ 500 \\ 600 \\ 700 \\ 800 \\ 900 \\ 1000 \\ 1100 \end{pmatrix} = \begin{pmatrix} .8414645 \\ .8413709 \\ .8413452 \\ .8413337 \\ .8413274 \\ .8413235 \\ .8413209 \\ .841319 \\ .8413176 \\ .8413165 \\ .8413156 \end{pmatrix}$$

and

$$
s\begin{pmatrix} 100 \\ 200 \\ 300 \\ 400 \\ 500 \\ 600 \\ 700 \\ 800 \\ 900 \\ 1000 \\ 1100 \end{pmatrix} = \begin{pmatrix} .84093\,78 \\ .84116\,18 \\ .84122\,34 \\ .84125\,08 \\ .84126\,58 \\ .84127\,52 \\ .84128\,15 \\ .84128\,6 \\ .84128\,94 \\ .84129\,2 \\ .84129\,41 \end{pmatrix}
$$

Be warned that these calculations are very intensive. Even if you have a fairly fast computer you may have to wait several minutes and if you don't have enough memory your computer may lock up. And notice how misleading these arrays are. The array of Simpson sums seems to indicate that $\int_0^1 \sqrt[3]{1-x^2}dx$ is .84129 to five decimal places. However, this isn't so. The Simpson sum with $n = 1100$ is good only to three decimal places and the corresponding midpoint sum is only slightly better.

6.5 A Convergent Improper Integral

In this section we consider the integral

$$
\int_0^1 \frac{(\sin x)\ln x}{x}dx. \tag{6.2}
$$

Since $\frac{\sin x}{x} \to 1$ as $x \to 0$ and since

$$
\lim_{\delta \to 0} \int_\delta^1 \ln x dx = \lim_{\delta \to 0} (\delta \ln \delta - \delta) = 0,
$$

The integral 6.2 converges. Clicking on Evaluate Numerically we obtain

$$
\int_0^1 \frac{(\sin x)\ln x}{x}dx = -.98181\,08.
$$

6.5.1 Approximating Sums for a Variety of Values of *n*

By giving *Scientific Notebook* the definitions

$$f(x) = \begin{cases} \frac{(\sin x)\ln x}{x} & \text{if } x \neq 0. \\ 0 & \text{if } x = 0. \end{cases},$$

$a = 0$ and $b = 1$, we can obtain the following table

	$n = 50$	$n = 100$	$n = 250$	$n = 500$
$L_f(a, b, n)$	$-.92428\,38$	$-.94958\,86$	$-.96709\,1$	$-.97375\,8$
$R_f(a, b, n)$	$-.92428\,38$	$-.94958\,86$	$-.96709\,1$	$-.97375\,8$
$T_f(a, b, n)$	$-.92428\,38$	$-.94958\,86$	$-.96709\,1$	$-.97375\,8$
$M_f(a, b, n)$	$-.97489\,33$	$-.97834\,86$	$-.98042\,51$	$-.98111\,78$
$S_f(a, b, n)$	$-.93885\,71$	$-.95802\,35$	$-.97107\,42$	$-.97598\,04$

Notice that the left sums, right sums and trapezoidal sums are exactly the same even though they are poor approximations to the integral. Can you explain why this is so? Notice that in this example too, the midpoint sums are more accurate than the Simpson sums.

6.6 Accuracy of the Sums

Looking at the preceding examples you will see that the Simpson sum is almost always a significantly more accurate approximation to the integral than are the left sum, right sum and trapezoidal sum. In some of the examples we saw that the midpoint sum is more accurate than the Simpson sum but this is not always the case. It can be shown, for example, that if f is a polynomial of degree 3 or less then the Simpson sums over an interval $[a, b]$ will be exactly equal to the integral $\int_a^b f(x)\,dx$.

6.7 An Indefinite Integral

It is easy to see that

$$\int \frac{1}{\sqrt{4 + x^2}}\,dx = \ln\left(x + \sqrt{(4 + x^2)}\right) + \text{constant.}$$

In this section we shall see how Riemann sums can be used to illustrate this fact. We begin by supplying *Scientific Notebook* with the definitions

$$f(x) = \frac{1}{\sqrt{4+x^2}}$$

and

$$F(x) = \int_0^x \frac{1}{\sqrt{2^2+t^2}}dt.$$

In order to illustrate the identity

$$F(x) = \ln\left(x + \sqrt{(4+x^2)}\right) - \ln 2,$$

we shall show that if x lies in the interval $[-4, 4]$ then the Riemann sum $M_f(0, x, 10)$ is a good approximation to $\ln\left(x + \sqrt{(4+x^2)}\right) - \ln 2$. Although we don't need to see an explicit formula for $M_f(0, x, 10)$ for this purpose, we could obtain one by pointing at $M_f(0, x, 10)$ and clicking on Evaluate. In fact,

$$\begin{aligned} M_f(0,x,10) &= \frac{2x}{\sqrt{(1600+x^2)}} + \frac{2x}{\sqrt{(1600+9x^2)}} + \frac{2x}{5\sqrt{(64+x^2)}} + \\ &\quad \frac{2x}{\sqrt{(1600+49x^2)}} + \frac{2x}{\sqrt{(1600+81x^2)}} + \frac{2x}{\sqrt{(1600+121x^2)}} + \\ &\quad \frac{2x}{\sqrt{(1600+169x^2)}} + \frac{2x}{5\sqrt{(64+9x^2)}} + \frac{2x}{\sqrt{(1600+289x^2)}} + \\ &\quad \frac{2x}{\sqrt{(1600+361x^2)}} + \frac{2x}{\sqrt{(1600+441x^2)}} \frac{2x}{\sqrt{(1600+441x^2)}}. \end{aligned}$$

However, even without this explicit formula we can point at the expression $M_f(0, x, 10)$ and click on Plot 2D and Rectangular, and then drag $\ln\left(x + \sqrt{(4+x^2)}\right) - \ln 2$ into the next figure to show how close the two graphs are to each other.

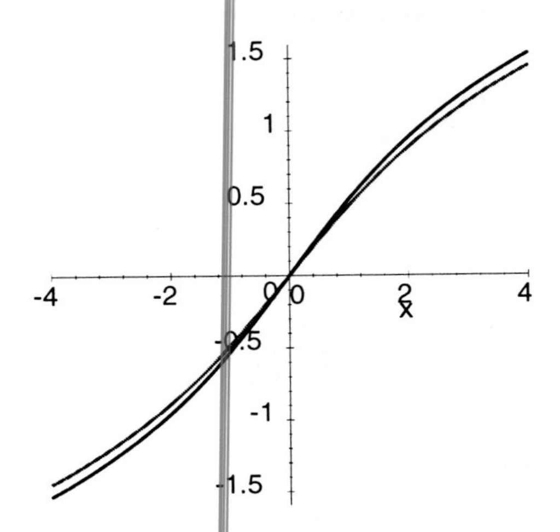

We can also demonstrate this closeness by showing that the area between the graphs of $M_f(0, x, n)$ and $\ln\left(x + \sqrt{(4 + x^2)}\right) - \ln 2$ becomes small as $n \to \infty$. We observe the following data:

$$\int_0^4 \left(M_f(0, x, 3) - \ln\left(x + \sqrt{(4 + x^2)}\right) + \ln 2\right) dx = .7598851$$

$$\int_0^4 \left(M_f(0, x, 5) - \ln\left(x + \sqrt{(4 + x^2)}\right) + \ln 2\right) dx = .4705763$$

$$\int_0^4 \left(M_f(0, x, 10) - \ln\left(x + \sqrt{(4 + x^2)}\right) + \ln 2\right) dx = .2411027$$

$$\int_0^4 \left(M_f(0, x, 20) - \ln\left(x + \sqrt{(4 + x^2)}\right) + \ln 2\right) dx = .1220601$$

$$\int_0^4 \left(M_f(0, x, 200) - \ln\left(x + \sqrt{(4 + x^2)}\right) + \ln 2\right) dx = 1.234504 \times 10^{-2}$$

6.8 The Fundamental Theorem of Calculus

In this section we shall use graphs of Riemann sums to provide an intuitive approach to the first form of the fundamental theorem of calculus. This theorem may be stated as follows:

6.8.1 Fundamental Theorem of Calculus (First Form)

Suppose that f is continuous on an open interval I. Suppose that a is any number in I and that the function F is defined by

$$F(x) = \int_a^x f(t)dt$$

for every $x \in I$. Then $F'(x) = f(x)$ for every number $x \in I$.

In the event that we are able to find an explicit formula for the expression $F(x) = \int_a^x f(t)dt$, we can check easily that the condition $F'(x) = f(x)$ holds. However, in almost every case we are unable produce a simple "closed form" expression for $F(x)$ by the process of taking limits of sums. In these cases we can still use *Scientific Notebook* to help us appreciate the fact that $F'(x) = f(x)$ for each x.

To illustrate the use of *Scientific Notebook* in this type of problem we define

$$f(x) = |1 - x \sin x|$$

for $x \in [-3, 3]$. What makes this function interesting is the fact that it fails to be differentiable at several points. Point at the latter equation and click on Define and New Definition. The graph of this function appears as follows:

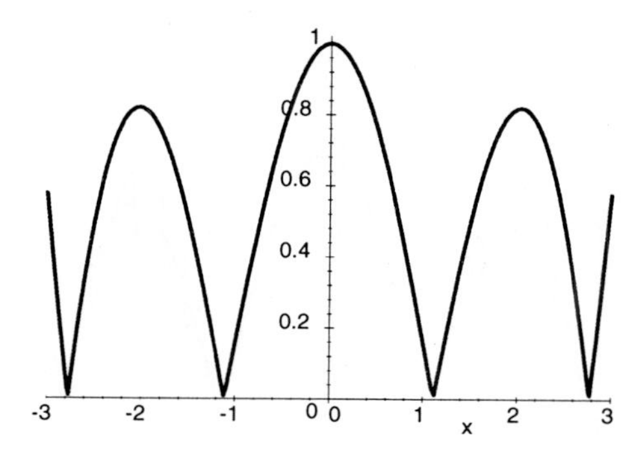

We shall now use midpoint sums sum to approximate $\int_0^x f(t)dt$. If $M_f(x,n)$ is the midpoint approximation to $\int_0^x f(t)\,dt$ taken over a regular partition of $[0,x]$ into n subdivisions then it is easy to see that

$$M_f(x,n) = \sum_{j=1}^{n} \left(\frac{x}{n}\right) f\left(\frac{2jx - x}{2n}\right).$$

Point at this equation and click on Define and New Definition.

As we know, if x is any number then

$$\lim_{n \to \infty} M_f(x,n) = \int_0^x f(t)dt.$$

To illustrate the fact that

$$\lim_{n \to \infty} M'_f(x, n) = f(x)$$

for each number x, we shall draw some graphs of the type $y = M'_f(x, n)$ for some chosen values of n and compare these graphs with the graph $y = f(x)$. Supply the definition $g(x) = M_f(x, 5)$ to *Scientific Notebook* by clicking on Define and New Definition and drag the expression $g'(x)$ into the graph of f. After adjusting the colors you will obtain the following figure:

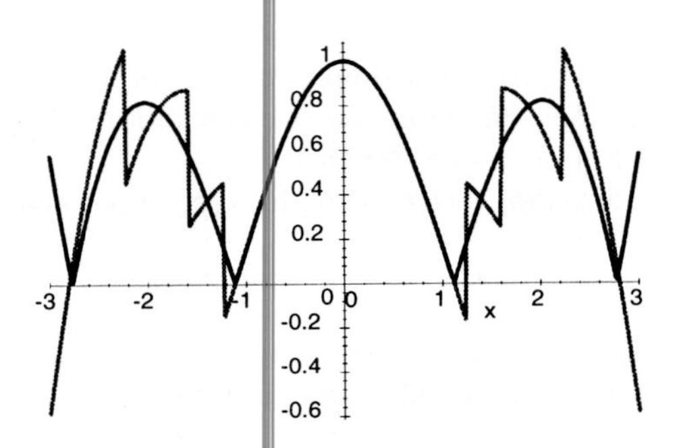

Note how the graphs of g' and f appear almost identical until the first point at which f fails to be differentiable. Repeat the procedure taking $g(x) = M_f(x, 5)$ and the figure becomes

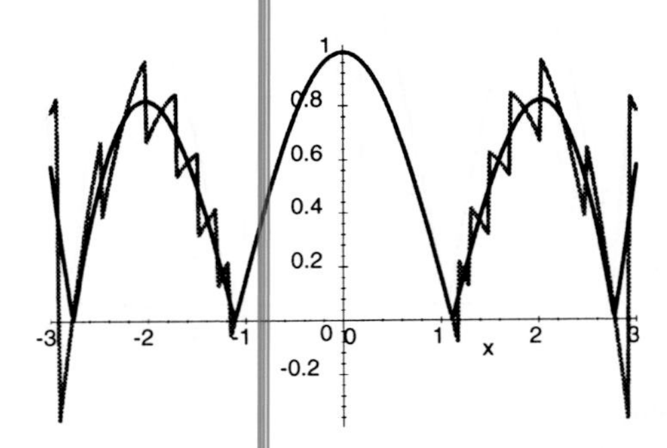

Try some further cases yourself. To see an animated illustration of the way the graphs $y = M'_f(x, n)$ approximate the graph $y = f(x)$ as n increases from 1 to 30, **click here**.

Chapter 7
Calculating Volumes

In this chapter we shall look at some standard volume problems that can be solved using single, double or triple integrals. We shall observe how *Scientific Notebook* can be used to simplify problems of this type. Not only can it evaluate the integrals for us, but it can also show us a sketch of the region whose volume we are finding.

7.1 Rotation of a Graph about an Axis

In this section we shall look at some regions that are generated when a curve is rotated about either the x-axis or the y-axis. We begin by defining a function g as follows:

$$g(x) = x\sin^2 x$$

for every number x and we inform *Scientific Notebook* of this definition by pointing at this equation and clicking on Define and New Definition. By pointing to the expression g and clicking on Plot 2D and Rectangular we can see the graph of g. If we restrict the domain of the function g to the interval $[0, \pi]$ then the graph appears as follows:

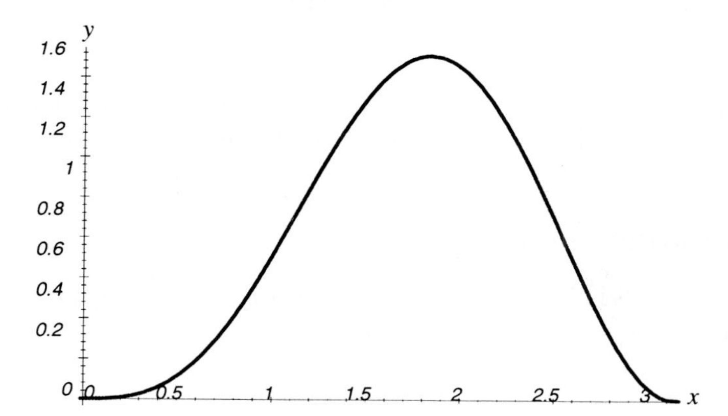

7.1.1 Rotating about the x-Axis

When the region that lies between the lines $x = 0$ and $x = \pi$, the graph $y = g(x)$ and the x-axis is rotated about the x-axis then the volume of the region so formed is

$$\int_0^{\pi} \pi \left(g\left(x\right)\right)^2 dx.$$

Since *Scientific Notebook* already knows the definition of the function g we can evaluate the integral in this form by pointing at it and clicking on Evaluate. We obtain

$$\int_0^{\pi} \pi (g(x))^2 dx = \frac{(8\pi^2 - 15)\pi^2}{64}.$$

If we want to sketch the solid region whose volume we have just found we can sketch the surface

$$y^2 + z^2 = x \sin^2 x$$

by pointing at its equation and clicking on Plot 3D and Implicit. We obtain the surface shown in the next figure:

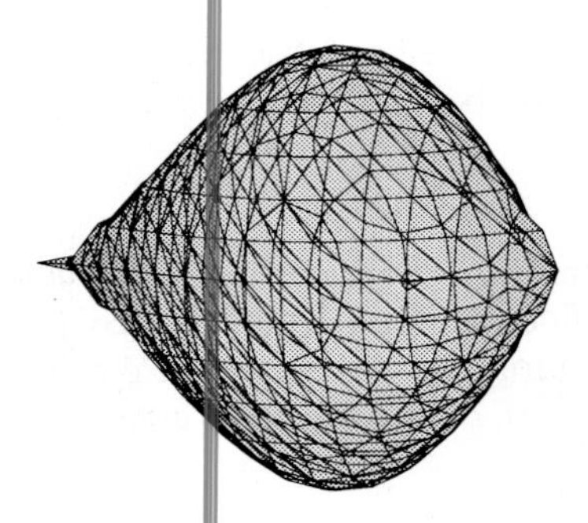

If your figure doesn't quite look like this, select it, click on the button at its lower right corner to bring up its properties dialog box and adjust the plot type, its ranges, its orientation and its sample size.

7.1.2 Rotating about the y-Axis

When the region that lies between the lines $x = 0$ and $x = \pi$, the graph $y = g(x)$ and the x-axis is rotated about the x-axis then the volume of the region so formed is

$$\int_0^{\pi} 2\pi x g(x) dx$$

and, pointing at this expression and clicking on Evaluate, we obtain

$$\int_0^{\pi} 2\pi x g(x) dx = \frac{\pi^2 (2\pi^2 - 3)}{6}.$$

To sketch a graph of the surface that is traced by the graph $y = g(x)$ as it is rotated about the y-axis we can ask *Scientific Notebook* to plot the graph of the parametric surface

$$x = t \cos \theta$$

$$\begin{aligned} y &= g(t) \\ z &= t\sin\theta \end{aligned}$$

where $\theta \in [0, 2\pi]$ and $t \in [0, \pi]$. For this purpose we point to the expression

$$[t\cos\theta, g(t), t\sin\theta]$$

and click on Plot 3D and Rectangular. Then we restrict θ and t to the desired intervals. In following figure we have allowed t to vary a little beyond π for visual effect.

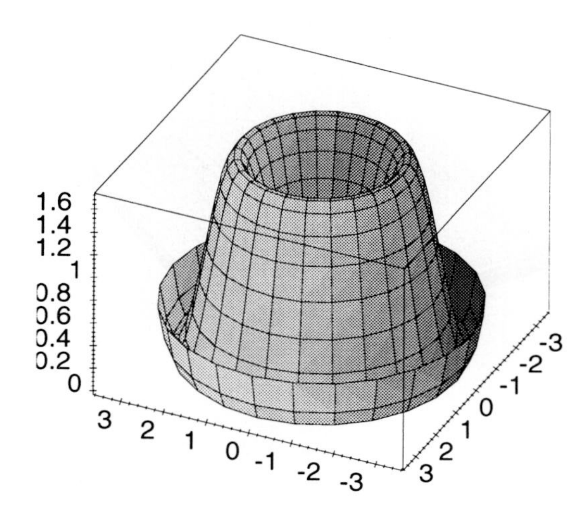

7.2 Applications of Double Integrals.

7.2.1 A Region Bounded by Planes

In this example we find the area of the region bounded by the planes

$$z = x + y, z = x + 7y, x = 0, x = 1, y = 1.$$

In order to sketch this region we begin by drawing the plane $z = x + y$. We write this plane in parametric form

$$\begin{aligned} x &= u \\ y &= v \\ z &= u + v \end{aligned}$$

where $0 \leq u \leq 1$ and $0 \leq v \leq 1$. Thus we can draw this plane by pointing at the expression

$(u, v, u+v)$, clicking on Plot 3D and Rectangular, and then adjusting the domain so that $0 \le u \le 1$ and $0 \le v \le 1$. To make the sketch more pleasing we suggest that you open the plot properties dialog box, select Patch for the Plot Style and xyz for the Shading. Once this part of the sketch has been done, highlight each of the expressions $(u, v, u+7v)$, $(0, u, v)$, $(1, u, v)$ and $(u, 1, v)$ in turn and drag it into the sketch. The sketch will now appear as follows:

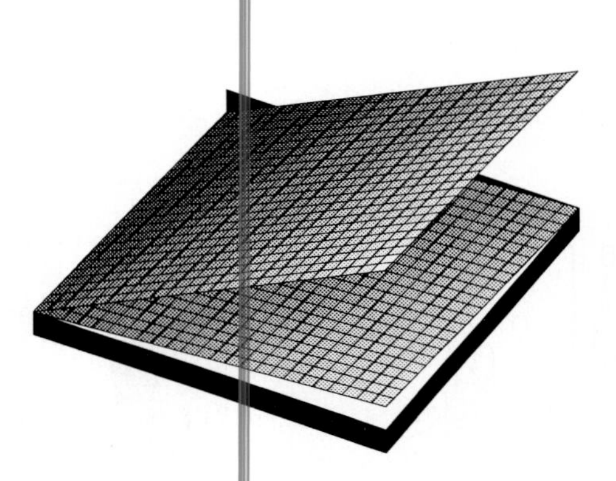

To find the volume, project the region onto the xy plane, the domain D is the square $[0,1] \times [0,1]$. The region is bounded above by $z = x + 7y$ and below by $z = x + y$. Thus the volume is

$$\int_0^1 \int_0^1 6y dy dx = 3.$$

7.2.2 A Region Bounded by Two Surfaces

In this example we find the volume of the region that lies between the surfaces $z = x^2 + 2y^2$ and $z = 6 - 2x^2 - y^2$. We see that these two surfaces intersect when

$$6 - 2x^2 - y^2 = x^2 + 2y^2,$$

in other words, $x^2 + y^2 = 2$. By pointing at the expression $6 - 2x^2 - y^2$, clicking on Plot 3D and Rectangular, and then dragging the expression $x^2 + 2y^2$ into this plot we obtain the sketch shown in the next figure. To make your figure resemble the one shown here, open the plot properties dialog box, go to the Plot Components screen and choose Patch for your plot style and xyz for your shading.

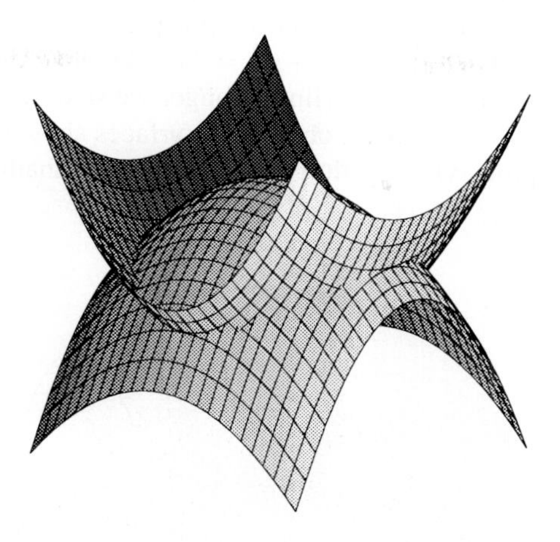

The volume of the region is therefore equal to

$$\iint_S (6 - 3x^2 - 3y^2)dydx$$

where

$$S = \{(x, y) \mid x^2 + y^2 \leq 2\}$$

This integral may be written in rectangular form as

$$\int_{-\sqrt{2}}^{\sqrt{2}} \int_{-\sqrt{2-x^2}}^{\sqrt{2-x^2}} (6 - 3x^2 - 3y^2)dydx$$

and by clicking on Evaluate we see that its value is 6π. Alternatively, we can write this integral in polar form as

$$\int_0^{2\pi} \int_0^{\sqrt{2}} \left(6 - 3r^2\right) rdrd\theta$$

and a click on Evaluate yields its value 6π.

7.2.3 Describing a Region with Cylindrical Polar Coordinates

In this section we discuss the region Ω that lies inside the cylinder $x^2 + y^2 = 2$, above the plane $z = 0$ and below the paraboloid $z = x^2 + y^2$.

Since the paraboloid is a set of points with cylindrical coordinates (r, θ, z) in which $z = r^2$ we can sketch this surface by pointing at the expression $\left(r, \theta, r^2\right)$ and clicking on Plot

3D and Cylindrical. The plane $z = 0$ can be described as a set of points with cylindrical coordinates of the form $(r, \theta, 0)$ and so we can add this plane to our sketch by dragging in the expression $(r, \theta, 0)$. Finally, the cylinder $x^2 + y^2 = 4$ can be thought of as a set of points with polar coordinates $(2, \theta, r)$. To make the cylinder longer we shall use $(2, \theta, 3r)$ and when this expression is dragged into the sketch we obtain the surfaces shown in the next figure. Once again we have selected a plot style of Patch and a xyz for the shading.

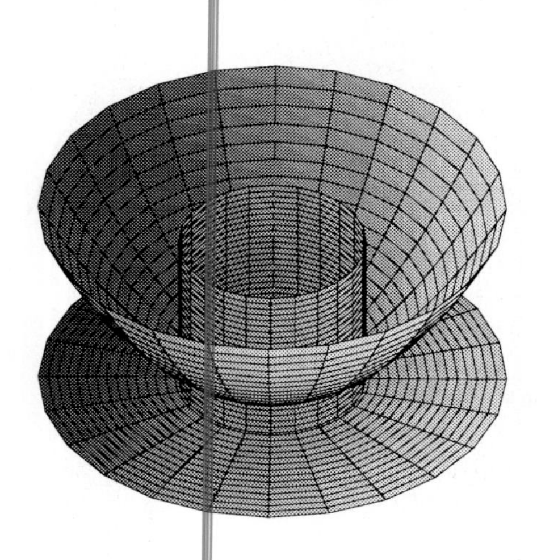

The volume of Ω can be expressed as

$$\iiint_{\Omega} 1 dx dy dz = \int_0^{2\pi} \int_0^2 \int_0^{r^2} 1 r dz dr d\theta = 8\pi.$$

Chapter 8
Linear Algebra with Scientific Workplace

In this chapter we explore some of the many operations of linear algebra that can be performed simply in *Scientific Notebook.*

8.1 Introduction

8.1.1 Using the Matrices Menu

Most of the linear algebra operations provided by *Scientific Notebook* can be performed by clicking on an item in the **matrices menu**. This appears as

Adjugate	Minimum Polynomial
Concatenate	Norm
Characteristic Polynomial	Nullspace Basis
Column Basis	Orthogonality Test
Condition Number	Permanent
Definiteness Tests	QR
Determinant	Rank
Eigenvalues	Rational Canonical Form
Eigenvectors	Reduced Row Echelon Form
Fill Matrix...	Reshape...
Fraction-Free Gaussian Elimination	Row Basis
Gaussian Elimination	Singular Values
Hermitian Transpose	SVD
Inverse	Smith Normal Form
Jordan Form	Trace
	Transpose

So, for example, one may find the inverse of the matrix

$$\begin{bmatrix} -19076342 & -23421223 & -50804332 & 23933017 \\ 22008670 & 24906060 & 59602522 & -23580148 \\ -3868770 & -98655 & -216016 & -6352815 \\ 8758630 & 13268746 & 26869750 & -17265686 \end{bmatrix}$$

by pointing at it, opening the matrices menu and clicking on Inverse.

This particular operation can also be performed without opening the matrices menu. We could simply point at the expression

$$\begin{bmatrix} -19076342 & -23421223 & -50804332 & 23933017 \\ 22008670 & 24906060 & 59602522 & -23580148 \\ -3868770 & -98655 & -216016 & -6352815 \\ 8758630 & 13268746 & 26869750 & -17265686 \end{bmatrix}^{-1}$$

and click on Evaluate. In either event, we obtain the inverse in the form

$$\begin{bmatrix} \frac{5792293}{212\,13864\,24004} & \frac{17021047}{1697\,10913\,92032} & -\frac{11798857}{848\,55456\,96016} & \frac{49669007}{1697\,10913\,92032} \\ -\frac{30997565}{848\,55456\,96016} & -\frac{3317043}{212\,13864\,24004} & \frac{11518081}{848\,55456\,96016} & -\frac{14542487}{424\,27728\,48008} \\ -\frac{245235}{212\,13864\,24004} & \frac{1135215}{1697\,10913\,92032} & \frac{1763519}{848\,55456\,96016} & -\frac{5567625}{1697\,10913\,92032} \\ -\frac{591085}{36\,89367\,69392} & -\frac{217219}{36\,89367\,69392} & \frac{243947}{36\,89367\,69392} & -\frac{633805}{36\,89367\,69392} \end{bmatrix}.$$

In the next subsection we provide some exercises that will help you to become familiar with some of the operations that are available from the matrices menu.

8.1.2 Some Exercises on Matrix Operations

1. Find the adjugate of the matrix

$$\begin{bmatrix} -19076342 & -23421223 & -50804332 & 23933017 \\ 22008670 & 24906060 & 59602522 & -23580148 \\ -3868770 & -98655 & -216016 & -6352815 \\ 8758630 & 13268746 & 26869750 & -17265686 \end{bmatrix}.$$

2. Given that

$$A = \begin{bmatrix} 2 & 2 & 3 & 1 \\ 1 & 3 & 5 & 7 \\ 1 & 1 & 2 & -3 \\ -2 & 1 & 1 & 3 \end{bmatrix}$$

and

$$B = \begin{bmatrix} 43 & -62 & 77 & 66 & 54 & -5 \\ 99 & -61 & -50 & -12 & -18 & 31 \\ -26 & -62 & 1 & -47 & -91 & -47 \\ -61 & 41 & -58 & -90 & 53 & -1 \end{bmatrix},$$

combine the matrices A and B to produce the matrix

$$\begin{bmatrix} 2 & 2 & 3 & 1 & 43 & -62 & 77 & 66 & 54 & -5 \\ 1 & 3 & 5 & 7 & 99 & -61 & -50 & -12 & -18 & 31 \\ 1 & 1 & 2 & -3 & -26 & -62 & 1 & -47 & -91 & -47 \\ -2 & 1 & 1 & 3 & -61 & 41 & -58 & -90 & 53 & -1 \end{bmatrix}.$$

Solution: Place the two matrices together and click on Concatenate.

3. Type the compound matrix

$$\begin{bmatrix} \begin{matrix} 1 & 2 \\ 3 & 4 \end{matrix} & 2 & 3 & & 1 \\ 1 & 3 & \begin{matrix} & 94 & \\ -85 & -55 & -37 \\ -35 & 97 & 50 \\ 79 & 56 & 49 \\ & -50 & \end{matrix} & \begin{matrix} 83 & -86 \\ -84 & 19 \\ 88 & -53 \end{matrix} & 7 \\ \begin{matrix} 1 \\ -2 \end{matrix} & \begin{matrix} 1 \\ 1 \end{matrix} & \begin{matrix} 2 \\ 1 \end{matrix} & & \begin{matrix} -3 \\ 3 \end{matrix} \end{bmatrix}.$$

4. Find a column basis for the matrix

$$A = \begin{matrix} 2 & 2 & 3 & -1 \\ 2 & 3 & 6 & 1 \\ 1 & 1 & 4 & 2 \\ -1 & 2 & 5 & 4 \end{matrix}.$$

5. Find a row basis for the matrix

$$A = \begin{bmatrix} 1 & 4 & 5 & 4 & 5 & 2 & 0 \\ 3 & 5 & -1 & 0 & 9 & 0 & 1 \\ 0 & -6 & -3 & 0 & 0 & 23 & 2 \\ 9 & 2 & -2 & 9 & 7 & 2 & 1 \\ 28 & 3 & 7 & 3 & 0 & -2 & -1 \\ 1 & 2 & 3 & 5 & 6 & 7 & 8 \end{bmatrix}$$

and the rank of this matrix.

6. Point at the equation

$$\begin{bmatrix} 2 & 2 & 3 & -1 \\ 2 & 3 & 6 & 1 \\ 1 & 1 & 4 & 2 \\ -1 & 2 & 5 & 4 \end{bmatrix} \begin{bmatrix} w \\ x \\ y \\ z \end{bmatrix} = \begin{bmatrix} 0 \\ 1 \\ 1 \\ 2 \end{bmatrix}$$

and click on Solve and Exact to obtain the solution in the form

$$\begin{bmatrix} -t_1 \\ \frac{1}{5} - t_1 \\ t_1 \\ \frac{2}{5} - t_1 \end{bmatrix}.$$

7. Point at the system of equations

$$\begin{aligned} 2w + 2x + 3y - z &= 0 \\ 2w + 3x + 6y + z &= 1 \\ w + x + 4y + 2z &= 1 \\ -w + 2x + 5y + 4z &= 2 \end{aligned}$$

and click on Solve and Exact. Compare the solution you have obtained with the solution obtained in the preceding exercise.

8. Find the reduced row echelon form of the matrix

$$\begin{bmatrix} 60 & 360 & 99 & 48 & 273 & 36 & 162 & 73 & -40 & 49 \\ 55 & 330 & 24 & 65 & -13 & 6 & 351 & -65 & 52 & 99 \\ -39 & -234 & 49 & -76 & 297 & -15 & -430 & 44 & -98 & 27 \\ 36 & 216 & 18 & 32 & 30 & 5 & 203 & 0 & 8 & 63 \\ -80 & -480 & -97 & 25 & -526 & -76 & -80 & -53 & -57 & -99 \\ 90 & 540 & 86 & -84 & 690 & 8 & 102 & -10 & 63 & -39 \\ -48 & -288 & -94 & 10 & -408 & -20 & -12 & 70 & 98 & 58 \end{bmatrix}.$$

9. Apply the process of Gaussian elimination to the matrix

$$\begin{bmatrix} 60 & 360 & 99 & 48 & 273 & 36 & 162 & 73 & -40 & 49 \\ 55 & 330 & 24 & 65 & -13 & 6 & 351 & -65 & 52 & 99 \\ -39 & -234 & 49 & -76 & 297 & -15 & -430 & 44 & -98 & 27 \\ 36 & 216 & 18 & 32 & 30 & 5 & 203 & 0 & 8 & 63 \\ -80 & -480 & -97 & 25 & -526 & -76 & -80 & -53 & -57 & -99 \\ 90 & 540 & 86 & -84 & 690 & 8 & 102 & -10 & 63 & -39 \\ -48 & -288 & -94 & 10 & -408 & -20 & -12 & 70 & 98 & 58 \end{bmatrix}.$$

10. Apply the process of fraction-free Gaussian elimination to the matrix

$$\begin{bmatrix} 60 & 360 & 99 & 48 & 273 & 36 & 162 & 73 & -40 & 49 \\ 55 & 330 & 24 & 65 & -13 & 6 & 351 & -65 & 52 & 99 \\ -39 & -234 & 49 & -76 & 297 & -15 & -430 & 44 & -98 & 27 \\ 36 & 216 & 18 & 32 & 30 & 5 & 203 & 0 & 8 & 63 \\ -80 & -480 & -97 & 25 & -526 & -76 & -80 & -53 & -57 & -99 \\ 90 & 540 & 86 & -84 & 690 & 8 & 102 & -10 & 63 & -39 \\ -48 & -288 & -94 & 10 & -408 & -20 & -12 & 70 & 98 & 58 \end{bmatrix}.$$

11. Given that

$$A = \begin{bmatrix} -19076342 & -23421223 & -50804332 & 23933017 \\ 22008670 & 24906060 & 59602522 & -23580148 \\ -3868770 & -98655 & -216016 & -6352815 \\ 8758630 & 13268746 & 26869750 & -17265686 \end{bmatrix},$$

find the value of e^A by giving *Scientific Notebook* the definition of A, pointing at e^A and clicking on Evaluate. Find an approximate value of e^A by clicking on Evaluate Numerically. Now evaluate $e^A e^{-A}$ and then verify that when you point at your answer and click on Simplify you obtain

$$\begin{bmatrix} 1 & 0 & 0 & 0 \\ 0 & 1 & 0 & 0 \\ 0 & 0 & 1 & 0 \\ 0 & 0 & 0 & 1 \end{bmatrix}.$$

12. Point at the equation $g(i, j) = i + j^2$ and click on Define and New Definition. Then open the matrices menu, click on Fill Matrix. Select six rows and eight columns, click on Defined by Function and fill in the function as g as shown.

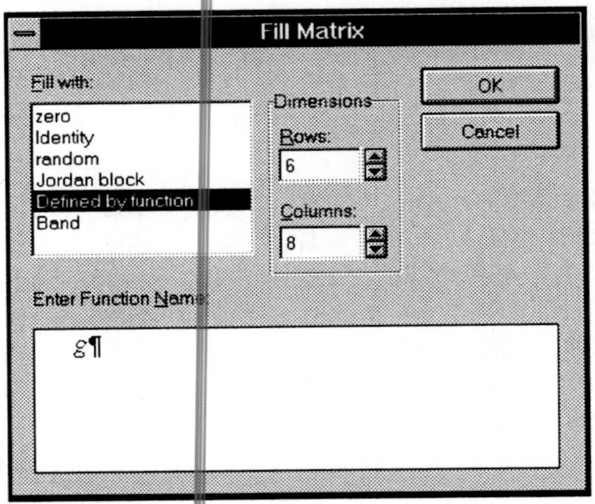

Fill Matrix

Fill with:
zero
Identity
random
Jordan block
Defined by function
Band

Dimensions
Rows: 6
Columns: 8

OK
Cancel

Enter Function Name:
g¶

After clicking on OK you should see the matrix

$$\begin{bmatrix} 2 & 5 & 10 & 17 & 26 & 37 & 50 & 65 \\ 3 & 6 & 11 & 18 & 27 & 38 & 51 & 66 \\ 4 & 7 & 12 & 19 & 28 & 39 & 52 & 67 \\ 5 & 8 & 13 & 20 & 29 & 40 & 53 & 68 \\ 6 & 9 & 14 & 21 & 30 & 41 & 54 & 69 \\ 7 & 10 & 15 & 22 & 31 & 42 & 55 & 70 \end{bmatrix}.$$

13. Use the Fill Matrix operation with the function $g(i,j) = x_i^{j-1}$ to produce the **Vandermonde matrix**

$$\begin{bmatrix} 1 & x_1 & x_1^2 & x_1^3 & x_1^4 & x_1^5 & x_1^6 & x_1^7 \\ 1 & x_2 & x_2^2 & x_2^3 & x_2^4 & x_2^5 & x_2^6 & x_2^7 \\ 1 & x_3 & x_3^2 & x_3^3 & x_3^4 & x_3^5 & x_3^6 & x_3^7 \\ 1 & x_4 & x_4^2 & x_4^3 & x_4^4 & x_4^5 & x_4^6 & x_4^7 \\ 1 & x_5 & x_5^2 & x_5^3 & x_5^4 & x_5^5 & x_5^6 & x_5^7 \\ 1 & x_6 & x_6^2 & x_6^3 & x_6^4 & x_6^5 & x_6^6 & x_6^7 \\ 1 & x_7 & x_7^2 & x_7^3 & x_7^4 & x_7^5 & x_7^6 & x_7^7 \\ 1 & x_8 & x_8^2 & x_8^3 & x_8^4 & x_8^5 & x_8^6 & x_8^7 \end{bmatrix}$$

14. Use the definition

$$a(n,k) = \begin{cases} \frac{2k}{n(n+1)} & \text{if } \; k \leq n \\ 0 & \text{if } \; k > n \end{cases}$$

and the Fill Matrix operation to create the matrix

$$\begin{bmatrix} 1 & 0 & 0 & 0 & 0 & 0 & 0 & 0 & 0 & 0 \\ \frac{1}{3} & \frac{2}{3} & 0 & 0 & 0 & 0 & 0 & 0 & 0 & 0 \\ \frac{1}{6} & \frac{1}{3} & \frac{1}{2} & 0 & 0 & 0 & 0 & 0 & 0 & 0 \\ \frac{1}{10} & \frac{1}{5} & \frac{3}{10} & \frac{2}{5} & 0 & 0 & 0 & 0 & 0 & 0 \\ \frac{1}{15} & \frac{2}{15} & \frac{1}{5} & \frac{4}{15} & \frac{1}{3} & 0 & 0 & 0 & 0 & 0 \\ \frac{1}{21} & \frac{2}{21} & \frac{1}{7} & \frac{4}{21} & \frac{5}{21} & \frac{2}{7} & 0 & 0 & 0 & 0 \\ \frac{1}{28} & \frac{1}{14} & \frac{3}{28} & \frac{1}{7} & \frac{5}{28} & \frac{3}{14} & \frac{1}{4} & 0 & 0 & 0 \\ \frac{1}{36} & \frac{1}{18} & \frac{1}{12} & \frac{1}{9} & \frac{5}{36} & \frac{1}{6} & \frac{7}{36} & \frac{2}{9} & 0 & 0 \\ \frac{1}{45} & \frac{2}{45} & \frac{1}{15} & \frac{4}{45} & \frac{1}{9} & \frac{2}{15} & \frac{7}{45} & \frac{8}{45} & \frac{1}{5} & 0 \\ \frac{1}{55} & \frac{2}{55} & \frac{3}{55} & \frac{4}{55} & \frac{1}{11} & \frac{6}{55} & \frac{7}{55} & \frac{8}{55} & \frac{9}{55} & \frac{2}{11} \end{bmatrix}$$

15. Point at the list of numbers

$$33, 35, 19, 29, 1, 3, 7, -2, 0, 1, 25, -8, 14,$$

open the matrices menu and click on Reshape. Fill in the number of columns as 3

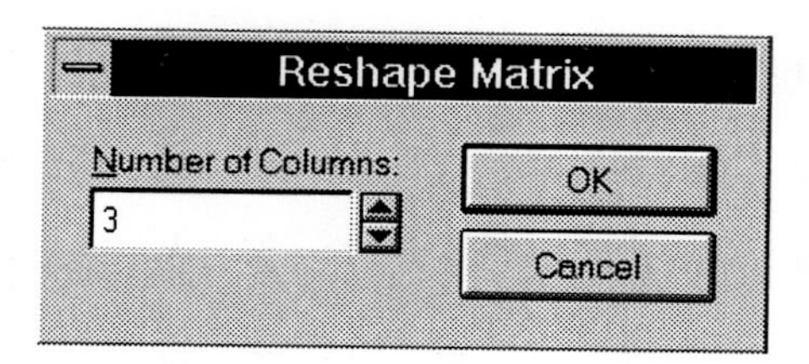

and obtain the matrix

$$\begin{bmatrix} 33 & 35 & 19 \\ 29 & 1 & 3 \\ 7 & -2 & 0 \\ 1 & 25 & -8 \\ 14 & & \end{bmatrix}.$$

Use a similar method to create the matrix

$$\begin{bmatrix} 33 & 35 & 19 & 29 & 1 & 3 \\ 7 & -2 & 0 & 1 & 25 & -8 \\ 14 & & & & & \end{bmatrix}.$$

16. Starting with the transpose of the matrix

$$\begin{bmatrix} 1 & 0 & 0 & 0 & 0 & 0 & 0 & 0 & 0 & 0 \\ \frac{1}{3} & \frac{2}{3} & 0 & 0 & 0 & 0 & 0 & 0 & 0 & 0 \\ \frac{1}{6} & \frac{1}{3} & \frac{1}{2} & 0 & 0 & 0 & 0 & 0 & 0 & 0 \\ \frac{1}{10} & \frac{1}{5} & \frac{3}{10} & \frac{2}{5} & 0 & 0 & 0 & 0 & 0 & 0 \\ \frac{1}{15} & \frac{2}{15} & \frac{1}{5} & \frac{4}{15} & \frac{1}{3} & 0 & 0 & 0 & 0 & 0 \\ \frac{1}{21} & \frac{2}{21} & \frac{1}{7} & \frac{4}{21} & \frac{5}{21} & \frac{2}{7} & 0 & 0 & 0 & 0 \\ \frac{1}{28} & \frac{1}{14} & \frac{3}{28} & \frac{1}{7} & \frac{5}{28} & \frac{3}{14} & \frac{1}{4} & 0 & 0 & 0 \\ \frac{1}{36} & \frac{1}{18} & \frac{1}{12} & \frac{1}{9} & \frac{5}{36} & \frac{1}{6} & \frac{7}{36} & \frac{2}{9} & 0 & 0 \\ \frac{1}{45} & \frac{2}{45} & \frac{1}{15} & \frac{4}{45} & \frac{1}{9} & \frac{2}{15} & \frac{7}{45} & \frac{8}{45} & \frac{1}{5} & 0 \\ \frac{1}{55} & \frac{2}{55} & \frac{3}{55} & \frac{4}{55} & \frac{1}{11} & \frac{6}{55} & \frac{7}{55} & \frac{8}{55} & \frac{9}{55} & \frac{2}{11} \end{bmatrix}$$

that you obtained in Exercise 14, use the Reshape operation to produce the matrix

$$\begin{bmatrix} 1 & 0 & 0 & 0 & 0 & \frac{1}{10} & 0 & \frac{1}{5} & 0 & \frac{5}{21} & \frac{1}{28} & \frac{1}{4} & \frac{1}{12} & 0 & \frac{1}{9} & \frac{1}{55} & \frac{7}{55} \\ 0 & 0 & 0 & 0 & 0 & \frac{1}{5} & 0 & \frac{4}{15} & 0 & \frac{2}{7} & \frac{1}{14} & 0 & \frac{1}{9} & 0 & \frac{2}{15} & \frac{2}{55} & \frac{8}{55} \\ 0 & 0 & 0 & \frac{1}{6} & 0 & \frac{3}{10} & 0 & \frac{1}{3} & \frac{1}{21} & 0 & \frac{3}{28} & 0 & \frac{5}{36} & \frac{1}{45} & \frac{7}{45} & \frac{3}{55} & \frac{9}{55} \\ 0 & 0 & 0 & \frac{1}{3} & 0 & \frac{2}{5} & 0 & 0 & \frac{2}{21} & 0 & \frac{1}{7} & 0 & \frac{1}{6} & \frac{2}{45} & \frac{8}{45} & \frac{4}{55} & \frac{2}{11} \\ 0 & \frac{1}{3} & 0 & \frac{1}{2} & 0 & 0 & \frac{1}{15} & 0 & \frac{1}{7} & 0 & \frac{5}{28} & \frac{1}{36} & \frac{7}{36} & \frac{1}{15} & \frac{1}{5} & \frac{1}{11} & \\ 0 & \frac{2}{3} & 0 & 0 & 0 & 0 & \frac{2}{15} & 0 & \frac{4}{21} & 0 & \frac{3}{14} & \frac{1}{18} & \frac{2}{9} & \frac{4}{45} & 0 & \frac{6}{55} & \end{bmatrix}.$$

8.2 Eigenvalues And Eigenvectors

8.2.1 Definition of an Eigenvalue

Suppose that A is a square matrix of order $n \times n$. A (real or complex) number λ is said to be an **eigenvalue** of the matrix A if the matrix $A - \lambda I$ is singular (where I stands for the identity matrix of order $n \times n$). Thus if λ is a given number then the following conditions are equivalent:

1. The number λ is an eigenvalue of A.
2. There exists a nonzero member x of $\mathbf{C}^n$ such that $(A - \lambda I)\, x = 0$.
3. The nullspace $N(A - \lambda I)$ of the matrix $A - \lambda I$ has positive dimension.
4. There exists a nonzero member x of $\mathbf{C}^n$ such that $Ax = \lambda x$.

8.2.2 Definition of an Eigenvector

Suppose that A is a square matrix of order $n \times n$ and that λ is an eigenvalue of A. Any nonzero member of $\mathbf{C}^n$ satisfying the equation $Ax = \lambda x$ is said to be an **eigenvector** of A **corresponding** to the eigenvalue λ.

8.2.3 The Minimal and Characteristic Polynomials

Suppose that A is a square matrix of order $n \times n$ and suppose that $\{\lambda_1, \lambda_2, \cdots, \lambda_k\}$ is the set of distinct eigenvalues of A. Given any $j \in \{1, \cdots, k\}$ we know that the nullspace $N(A - \lambda_j I)$ of the matrix $A - \lambda_j I$ is not just $\{0\}$. Furthermore,

$$N(A - \lambda_j I) \subseteq N\left((A - \lambda_j I)^2\right) \subseteq N\left((A - \lambda_j I)^3\right) \subseteq \cdots.$$

This expanding sequence of subspaces of $\mathbf{C}^n$ must eventually become constant. We define p_j to be the least positive integer r for which

$$N\left((A - \lambda_j I)^r\right) = N\left((A - \lambda_j I)^{r+1}\right)$$

and we define q_j to be the dimension of the space $N\left((A - \lambda_j I)^{p_j}\right)$. Observe that $p_j \leq q_j$.

The **minimal polynomial** and **characteristic polynomial** of the matrix A are the polynomials f and g defined by

$$f(\lambda) = (\lambda - \lambda_1)^{p_1} (\lambda - \lambda_2)^{p_2} \cdots (\lambda - \lambda_k)^{p_k}$$

and

$$g(\lambda) = (\lambda - \lambda_1)^{q_1} (\lambda - \lambda_2)^{q_2} \cdots (\lambda - \lambda_k)^{q_k}$$

for every number λ. Some of the main facts about these polynomials are listed below:

1. The minimal polynomial is a factor of the characteristic polynomial..
2. Both $f(A)$ and $g(A)$ are equal to the zero matrix. This fact is known as the Cayley–Hamilton theorem.
3. The space $\mathbf{C}^n$ is the direct sum of the spaces $N\left((A - \lambda_j I)^{p_j}\right)$ as j runs from 1 to k and we have $\sum_{j=1}^{k} q_j = n$
4. The value $g(\lambda)$ of the characteristic polynomial is equal to the determinant of the matrix $A - \lambda I$.

8.2.4 Finding Eigenvalues and Eigenvectors

1. Find the eigenvalues of the matrix

$$\begin{bmatrix} -19076342 & -23421223 & -50804332 & 23933017 \\ 22008670 & 24906060 & 59602522 & -23580148 \\ -3868770 & -98655 & -216016 & -6352815 \\ 8758630 & 13268746 & 26869750 & -17265686 \end{bmatrix}$$

by pointing at it, opening the matrices menu and clicking on Eigenvalues. This operation yields. 2912996, −2912996, −5825992, −5825992.

2. Find the eigenvectors of the matrix

$$\begin{bmatrix} -19076342 & -23421223 & -50804332 & 23933017 \\ 22008670 & 24906060 & 59602522 & -23580148 \\ -3868770 & -98655 & -216016 & -6352815 \\ 8758630 & 13268746 & 26869750 & -17265686 \end{bmatrix}$$

by clicking on Eigenvectors. The eigenvectors appear matched with their corresponding eigenvalues as shown.

$$\left\{ \begin{bmatrix} -\frac{133337}{68264} \\ \frac{74049}{34132} \\ \frac{3085}{9752} \\ 1 \end{bmatrix} \right\} \leftrightarrow 2912996$$

$$\left\{ \begin{bmatrix} \frac{367657}{11217} \\ -\frac{494140}{11217} \\ 1 \\ -\frac{211462}{11217} \end{bmatrix} \right\} \leftrightarrow -2912996$$

$$\left\{ \begin{bmatrix} -\frac{1068}{631} \\ \frac{1249}{631} \\ 0 \\ 1 \end{bmatrix}, \begin{bmatrix} \frac{4819}{3155} \\ -\frac{1914}{631} \\ 1 \\ 0 \end{bmatrix} \right\} \leftrightarrow -5825992$$

Observe that the eigenspace corresponding to the eigenvalue −5825992 has dimension 2.

3. Find the characteristic and minimal polynomials of the matrix

$$\begin{bmatrix} -19076342 & -23421223 & -50804332 & 23933017 \\ 22008670 & 24906060 & 59602522 & -23580148 \\ -3868770 & -98655 & -216016 & -6352815 \\ 8758630 & 13268746 & 26869750 & -17265686 \end{bmatrix}$$

by pointing at the matrix and clicking on either Characteristic Polynomial or Minimal Polynomial. Observe that the two polynomials are not the same. As a matter of fact, you can use the preceding exercise to see that the minimal polynomial of this matrix is

$$(\lambda - 2912996)(\lambda + 2912996)(\lambda + 5825992)$$

and that the characteristic polynomial is

$$(\lambda - 2912996)(\lambda + 2912996)(\lambda + 5825992)^2 .$$

4. In this exercise we explore the eigenvalues of a matrix that have to be found numerically. Point to the matrix

$$\begin{bmatrix} 43 & -62 & 77 & 66 & 54 & -5 \\ 99 & -61 & -50 & -12 & -18 & 31 \\ -26 & -62 & 1 & -47 & -91 & -47 \\ -61 & 41 & -58 & -90 & 53 & -1 \\ 94 & 83 & -86 & 23 & -84 & 19 \\ -50 & 88 & -53 & 85 & 49 & 78 \end{bmatrix}$$

and click on Eigenvalues. You will obtain:
eigenvalues: ρ where ρ is a root of

$$6216374\,99712 + 1\,10075\,88903Z - 136540544Z^2 - 3736896Z^3 - 19131Z^4 + 113Z^5 + Z^6.$$

Scientific Notebook has provided you with the characteristic polynomial. Now point at this polynomial and click on Polynomials and Roots and you will obtain the list

$$\begin{array}{c} -120.05 \\ -76.534 - 61.781i \\ -76.534 + 61.781i \\ -58.06 \\ 57.3 \\ 160.88 \end{array}$$

of eigenvalues.

5. In this exercise we repeat the preceding exercise using a slightly different method. We convert one or more of the entries of the given matrix to a decimal form in order to force *Scientific Notebook* to work numerically when we request the eigenvalues. Point at the

matrix

$$\begin{bmatrix} 43.0 & -62 & 77 & 66 & 54 & -5 \\ 99 & -61 & -50 & -12 & -18 & 31 \\ -26 & -62 & 1 & -47 & -91 & -47 \\ -61 & 41 & -58 & -90 & 53 & -1 \\ 94 & 83 & -86 & 23 & -84 & 19 \\ -50 & 88 & -53 & 85 & 49 & 78 \end{bmatrix}$$

and click on **Eigenvalues**. $160.88, 57.3, -76.534 + 61.781i, -76.534 - 61.781i, -58.06, -120.05$

6. In this exercise we explore the eigenvectors of a matrix that have to be found numerically. Point to the matrix

$$\begin{bmatrix} 43 & -62 & 77 & 66 & 54 & -5 \\ 99 & -61 & -50 & -12 & -18 & 31 \\ -26 & -62 & 1 & -47 & -91 & -47 \\ -61 & 41 & -58 & -90 & 53 & -1 \\ 94 & 83 & -86 & 23 & -84 & 19 \\ -50 & 88 & -53 & 85 & 49 & 78 \end{bmatrix}$$

and click on **Eigenvectors**. You will obtain[9]

$$\left\{ \left[\begin{matrix} -\frac{33749\,44578\,46499\,84287\,19358}{1164\,87243\,92556\,99689\,10624\,71819}\rho^2 + \frac{7\,45932\,76284\,44294\,38796}{1164\,87243\,92556\,99689\,10624\,71819}\rho^3 + \frac{1\,37961\,06734\,3508}{1164\,87243\,92556\,99689} \\ \frac{355\,28485\,22082\,38492\,22790}{12813\,59683\,18126\,96580\,16871\,90009}\rho^3 + \frac{28104\,45079\,45654\,05585}{25627\,19366\,36253\,93160\,33743\,80018}\rho^4 + \frac{4112\,29461\,26256\,2}{25627\,19366\,36253\,93160} \\ \\ -\frac{6\,02405\,60706\,06987\,45581\,44051}{51254\,38732\,72507\,86320\,67487\,60036}\rho^2 - \frac{3037\,35442\,28665\,06253\,75017\,48177}{12813\,59683\,18126\,96580\,16871\,90009} - \frac{11891\,56706\,05286\,89048}{51254\,38732\,72507\,86320\,67} \\ -\frac{1\,19088\,16326\,06742\,59884\,68569}{51254\,38732\,72507\,86320\,67487\,60036}\rho^2 + \frac{3157\,05534\,02973\,65192\,78791}{51254\,38732\,72507\,86320\,67487\,60036}\rho^3 + \frac{6\,14261\,36538\,5903}{12813\,59683\,18126\,96580} \\ \frac{930\,67820\,06553\,44329\,04521}{12813\,59683\,18126\,96580\,16871\,90009}\rho^3 + \frac{583\,80023\,98739\,19706\,24039\,04916}{12813\,59683\,18126\,96580\,16871\,90009}\rho - \frac{20\,47194\,98537\,91072}{12813\,59683\,18126\,96580\,16} \end{matrix} \right. \right.$$

$$\rho$$

where ρ is a root of

$$62\,16374\,99712 + 1\,10075\,88903Z - 136540544Z^2 - 3736896Z^3 - 19131Z^4 + 113Z^5 + Z^6.$$

Point at this expression and click on **Polynomials** and **Roots**. You will obtain

$$\begin{matrix} -120.05 \\ -76.534 - 61.781i \\ -76.534 + 61.781i \\ -58.06 \\ 57.3 \\ 160.88 \end{matrix}$$

---

[9] The following expression will be fully visible only in the on-screen version of this text.

Highlight the equation $\rho = -120.05$ and click on Define and New Definition. Then drag down a copy[10] of the above formula

$$\begin{bmatrix}
-\frac{33749\,44578\,46499\,84287\,19358}{1164\,87243\,92556\,99689\,10624\,71819}\rho^2 + \frac{7\,45932\,76284\,44294\,38796}{1164\,87243\,92556\,99689\,10624\,71819}\rho^3 + \frac{1\,37961\,06734\,35082}{1164\,87243\,92556\,99689\,1} \\
\frac{355\,28485\,22082\,38492\,22790}{12813\,59683\,18126\,96580\,16871\,90009}\rho^3 + \frac{28104\,45079\,45654\,05585}{25627\,19366\,36253\,93160\,33743\,80018}\rho^4 + \frac{4112\,29461\,26256\,251}{25627\,19366\,36253\,93160\,337} \\
\\
-\frac{6\,02405\,60706\,06987\,45581\,44051}{51254\,38732\,72507\,86320\,67487\,60036}\rho^2 - \frac{3037\,35442\,28665\,06253\,75017\,48177}{12813\,59683\,18126\,96580\,16871\,90009} - \frac{11891\,56706\,05286\,89048\,0}{51254\,38732\,72507\,86320\,6748} \\
-\frac{1\,19088\,16326\,06742\,59884\,68569}{51254\,38732\,72507\,86320\,67487\,60036}\rho^2 + \frac{3157\,05534\,02973\,65192\,78791}{51254\,38732\,72507\,86320\,67487\,60036}\rho^3 + \frac{6\,14261\,36538\,59032\,0}{12813\,59683\,18126\,96580\,16} \\
\frac{930\,67820\,06553\,44329\,04521}{12813\,59683\,18126\,96580\,16871\,90009}\rho^3 + \frac{583\,80023\,98739\,19706\,24039\,04916}{12813\,59683\,18126\,96580\,16871\,90009}\rho - \frac{20\,47194\,98537\,91072\,325}{12813\,59683\,18126\,96580\,1687}
\end{bmatrix}$$

for an eigenvector corresponding to ρ, highlight this formula, hold down the control key and click on Evaluate Numerically. You will obtain.

$$\begin{bmatrix} 2.727263 \times 10^{-2} \\ .7446578 \\ 1.0 \\ -1.799373 \\ 1.576115 \\ .325928 \end{bmatrix}$$

Repeat this process for each of the other five eigenvectors and you will obtain the eigenvectors

$$\begin{bmatrix} 2.727263 \times 10^{-2} \\ .7446578 \\ 1.0 \\ -1.799373 \\ 1.576115 \\ .325928 \end{bmatrix} \begin{bmatrix} .3177171 \\ .9006995 \\ 1.0 \\ -.9410832 \\ .2043806 \\ .4380539 \end{bmatrix} \begin{bmatrix} -2.690259 \\ -.7984595 \\ 1.0 \\ -.3107667 \\ -2.138673 \\ 5.795245 \end{bmatrix} \begin{bmatrix} .3899299 \\ -.2528862 \\ 1.0 \\ -.4606816 \\ -.477767 \\ -1.89813 \end{bmatrix}$$

$$\begin{bmatrix} .48434 + .58142i \\ 1.5511 - .48943i \\ 1.0 \\ -1.5172 - .3446i \\ .46716 - .79155i \\ -5.1719 \times 10^{-2} + .88668i \end{bmatrix} \begin{bmatrix} .48434 - .58142i \\ 1.5511 + .48943i \\ 1.0 \\ -1.5172 + .3446i \\ .46716 + .79155i \\ -5.1719 \times 10^{-2} - .88668i \end{bmatrix}$$

[10] Again, the entire expression will be visible only in the on-screen version of this text.

7. We now repeat the preceding exercise working with the matrix

$$A = \begin{bmatrix} 43.0 & -62 & 77 & 66 & 54 & -5 \\ 99 & -61 & -50 & -12 & -18 & 31 \\ -26 & -62 & 1 & -47 & -91 & -47 \\ -61 & 41 & -58 & -90 & 53 & -1 \\ 94 & 83 & -86 & 23 & -84 & 19 \\ -50 & 88 & -53 & 85 & 49 & 78 \end{bmatrix}$$

which is obtained by converting an entry of the given matrix to decimal form. As we have said, this form of the matrix A forces Scientific Notebook to work numerically instead of trying to give exact solutions. Point at the matrix A defined in this way and click on Eigenvectors. You will obtain

$$\left\{ \begin{bmatrix} .18252 \\ -.11837 \\ .46808 \\ -.21564 \\ -.22364 \\ -.88848 \end{bmatrix} \right\} \leftrightarrow 160.88$$

$$\left\{ \begin{bmatrix} -.77086 \\ -.22879 \\ .28654 \\ -8.9047 \times 10^{-2} \\ -.61281 \\ 1.6606 \end{bmatrix} \right\} \leftrightarrow 57.3$$

$$\left\{ \begin{bmatrix} -.40636 + 7.3199 \times 10^{-2} i \\ -.18559 + .86785i \\ -.26937 + .47451i \\ .57221 - .62707i \\ .24976 + .43489i \\ -.40681 - .2634i \end{bmatrix} \right\} \leftrightarrow -76.534 + 61.781i$$

$$\left\{ \begin{bmatrix} -.40636 - 7.3199 \times 10^{-2} i \\ -.18559 - .86785i \\ -.26937 - .47451i \\ .57221 + .62707i \\ .24976 - .43489i \\ -.40681 + .2634i \end{bmatrix} \right\} \leftrightarrow -76.534 - 61.781i$$

$$\left\{\begin{bmatrix} .73961 \\ 2.0967 \\ 2.3279 \\ -2.1907 \\ .47578 \\ 1.0197 \end{bmatrix}\right\} \leftrightarrow -58.06$$

$$\left\{\begin{bmatrix} 1.5258 \times 10^{-2} \\ .41661 \\ .55947 \\ -1.0067 \\ .88179 \\ .18235 \end{bmatrix}\right\} \leftrightarrow -120.05$$

Notice how much quicker the process is this way.

8.2.5 Exploring the Eigenspaces of a Matrix

In this subsection we shall investigate the matrix

$$A = \begin{bmatrix} 1799 & 2532 & 3069 & 3536 & 4031 & 991 & -1118 \\ 147 & 12 & 269 & 260 & 279 & 291 & 254 \\ 175 & 236 & 101 & 92 & 307 & 319 & 282 \\ 2163 & 2672 & 3181 & 1856 & 951 & 1411 & 1822 \\ 1967 & 2784 & 3601 & 6112 & 8063 & 599 & -446 \\ -3689 & -4300 & -4911 & -5592 & -6245 & 4855 & 2578 \\ -1729 & -2676 & -3623 & -4640 & -5629 & -6625 & -1790 \end{bmatrix}.$$

Before we begin, we point at this definition of A and click on Define and New Definition. Then we point at the equation

$$I = \begin{bmatrix} 1 & 0 & 0 & 0 & 0 & 0 & 0 \\ 0 & 1 & 0 & 0 & 0 & 0 & 0 \\ 0 & 0 & 1 & 0 & 0 & 0 & 0 \\ 0 & 0 & 0 & 1 & 0 & 0 & 0 \\ 0 & 0 & 0 & 0 & 1 & 0 & 0 \\ 0 & 0 & 0 & 0 & 0 & 1 & 0 \\ 0 & 0 & 0 & 0 & 0 & 0 & 1 \end{bmatrix}$$

and click on Define and New Definition. We now explore the matrix A as follows.

8.2.5.1 Find the Eigenvalues

In this step we find and name the eigenvalues of A. Point at the matrix A and click on Eigenvalues. You will obtain $-196, -196, 588, 588, 4704, 4704, 4704$. Thus the set of eigenvalues of A is $\{-196, 588, 4704\}$. Point at each of the equations $\lambda_1 = -196$, $\lambda_2 = 588$ and $\lambda_3 = 4704$ and click on Define and New Definition.

8.2.5.2 The First Sequence of Eigenspaces

In this step we study the successive nullspaces of the matrix $A - \lambda_1 I$. Point at the matrix $A - \lambda_1 I$, and click on Nullspace Basis. You will obtain

$$\{(1, -2, 1, 0, 0, 0, 0), (28, -25, 0, 1, 1, 1, 1)\}$$

thus showing that

$$\dim(\mathcal{N}(A - \lambda_1 I)) = 2.$$

Now point at the matrix $(A - \lambda_1 I)^2$ and click on Nullspace Basis. Once again you will obtain

$$\{(1, -2, 1, 0, 0, 0, 0), (28, -25, 0, 1, 1, 1, 1)\}$$

and, of course,

$$\dim\left(\mathcal{N}(A - \lambda_1 I)^2\right) = 2.$$

This should come as no surprise because the multiplicity of the eigenvalue λ_1 is only 2. Thus, according to the notation introduced in Subsection 8.2.3 we have $p_1 = 1$ and $q_1 = 2$.

8.2.5.3 The Second Sequence of Eigenspaces

Point at the matrix $A - \lambda_2 I$, and click on Nullspace Basis. You will obtain

$$\{(1, 1, -13, 57, -41, 1, 1)\}$$

thus showing that

$$\dim(\mathcal{N}(A - \lambda_2 I)) = 1.$$

Now point at the matrix $(A - \lambda_2 I)^2$ and click on Nullspace Basis. You will obtain

$$\{(0, 0, 1, -2, 1, 0, 0), (1, 1, 28, -25, 0, 1, 1)\}$$

and so

$$\dim\left(\mathcal{N}(A - \lambda_2 I)^2\right) = 2.$$

Now point to the matrix $(A - \lambda_2 I)^3$ and click on Nullspace Basis. You will obtain

$$\{(0, 0, 1, -2, 1, 0, 0), (1, 1, 28, -25, 0, 1, 1)\}$$

again. Once again, this is not surprising since the multiplicity of the eigenvalue λ_2 is only 2. We see, therefore, that $p_2 = q_2 = 2$.

8.2.5.4 The Third Sequence of Eigenspaces

Point at the matrix $A - \lambda_3 I$, and click on Nullspace Basis. You will obtain

$$\{(29, 1, 1, 1, -83, 225, -167)\}$$

and so

$$\dim (N (A - \lambda_2 I)) = 1.$$

Now point at the matrix $(A - \lambda_3 I)^2$ and click on Nullspace Basis. You will obtain

$$\{(-1, 0, 0, 0, 1, -3, 3), (-54, 1, 1, 1, 0, -24, 82)\}$$

and so

$$\dim \left(N (A - \lambda_2 I)^2 \right) = 2.$$

Now point at the matrix $(A - \lambda_3 I)^3$ and click on Nullspace Basis. You will obtain

$$\{(2, 0, 0, 0, 1, 0, -3), (1, 0, 0, 0, 0, 1, -2), (-30, 1, 1, 1, 0, 0, 34)\}$$

and so

$$\dim \left(N (A - \lambda_2 I)^3 \right) = 3.$$

Finally, point at the matrix $(A - \lambda_3 I)^4$ and click on Nullspace Basis and you will see that

$$\dim \left(N (A - \lambda_2 I)^4 \right) = 3.$$

We see, therefore, that $p_3 = q_3 = 3$.

8.2.5.5 The Minimal and Characteristic Polynomials

If f and g are the minimal and characteristic polynomials of A then we have $\lambda_1 = -196$, $\lambda_2 = 588$ and $\lambda_3 = 4704$

$$f (\lambda) = (\lambda + 196) (\lambda - 588)^2 (\lambda - 4704)^3$$

and

$$g (\lambda) = (\lambda + 196)^2 (\lambda - 588)^2 (\lambda - 4704)^3$$

for every number λ. By clicking directly on Minimal Polynomial and Characteristic Polynomial we can verify these answers.

8.2.6 Exercise on Eigenspaces

Discuss the eigenspaces of the 13×13 matrix[11]

[11] The following matrix will be fully visible only in the on-screen version of this text.

$$\begin{bmatrix} 2637876 & 709388 & 1747610 & 738668 & -3004638 & -818550 & 1337152 \\ -7309352 & -3483506 & -3236540 & -2444960 & 10041822 & 3401010 & -2624570 \\ 3312412 & 952396 & 1621986 & 876372 & -2921114 & -1436658 & 995042 \\ 10572206 & 5207172 & 4897500 & 3679438 & -15447642 & -4597520 & 4127512 \\ -7304922 & -3708820 & -3192264 & -2668876 & 10094175 & 3196691 & -2851619 \\ -29702724 & -14168626 & -12905800 & -10251514 & 39303901 & 13213019 & -9770209 \\ -2548652 & -1314430 & -1016882 & -1132348 & 2297120 & 639510 & -489444 \\ 8185856 & 3970848 & 2523402 & 2681308 & -9895564 & -4495358 & 1314838 \\ -21234890 & -9466146 & -9806484 & -7013804 & 28479269 & 9477691 & -7696793 \\ 23763608 & 11737486 & 9713948 & 8302448 & -31193918 & -10861036 & 7568152 \\ -426800 & -806002 & -119886 & -320572 & 2406586 & 387168 & -467346 \\ 3843874 & 2258538 & 908622 & 1425088 & -4942665 & -2352941 & 229029 \\ 6365274 & 2853678 & 2939980 & 2262148 & -7689969 & -2344023 & 2202805 \end{bmatrix}$$

The fact that this matrix doesn't fit onto your screen does not present a problem. If you like, you can scroll horizontally to see the part that is presently hidden. On the other hand, if you give this matrix the name A, supply this definition to *Scientific Notebook* and work with the symbol A then you can work without having to scroll at all.

8.3 Norm of a Matrix

In this section we shall show how to calculate the norm of any given square matrix numerically with *Scientific Notebook*. The quickest way to find the norm of a given matrix A is to rewrite it in such a way that one or more of its entries is in decimal form (for example 4.0 instead of 4) and then click on Matrices and Norm. This method can be used to check the accuracy of the method we describe below. However, the method described below is meant to do more than merely find the norm. It is meant to teach you what the norm of a matrix is and how we should go about calculating it. We begin with the definition of this norm.

8.3.1 Defining the Norm

Given a point $x = (x_1, x_2, \cdots, x_n) \in \mathbf{R}^n$, we shall use the symbol $\|x\|$ for the 2-norm $\|x\|_2$. In other words,

$$\|x\| = \sqrt{\sum_{i=1}^{n} x_i^2}\,.$$

If A is a matrix of order $n \times n$, then the norm $\|A\|$ of the matrix A is defined to be the norm of the linear operator $L_A : \mathbf{R}^n \to \mathbf{R}^n$ which is defined by

$$L_A(x) = Ax$$

for every point $x \in \mathbf{R}^n$. In other words,

$$\|A\| = \max\left\{\|Ax\| \mid x \in \mathbf{R}^n \text{ and } \|x\| = 1\right\}.$$

8.3.2 The Quickest Way to Find a Norm

Given a square matrix A, the quickest way to calculate an approximation to $\|A\|$ is to rewrite the matrix with one or more of its entries in decimal form. This forces *Scientific Notebook* to work with the matrix numerically instead of exactly. Thus, in order to find the norm of the matrix

$$A = \begin{bmatrix} -1 & -7 & -1 & 4 & -5 \\ -9 & 5 & -1 & 4 & 3 \\ -8 & 3 & -4 & 9 & -5 \\ 8 & -5 & 8 & 9 & 7 \\ 1 & 2 & -9 & -5 & -8 \end{bmatrix}$$

We write it in the form

$$A = \begin{bmatrix} -1 & -7 & -1 & 4 & -5 \\ -9 & 5 & -1 & 4.0 & 3 \\ -8 & 3 & -4 & 9 & -5 \\ 8 & -5 & 8 & 9 & 7 \\ 1 & 2 & -9 & -5 & -8 \end{bmatrix}$$

and supply *Scientific Notebook* with this definition. Then we point at the matrix A and click on Norm. This yields 2-norm: 21.027. Alternatively, we can point at the expression $\|A\|$ and click on Evaluate. This gives us $\|A\| = 21.027$.

8.3.3 A Procedure for Calculating the Norm of a Matrix

Suppose that $A = [a_{ij}]$ is a square matrix of order $n \times n$. In order to maximize the expression $\|Ax\|^2$ subject to the constraint $\|x\|^2 = 1$, we shall use the method of Lagrange multipliers. This method requires that we solve the system of equations

$$\begin{aligned} \nabla \|Ax\|^2 - \lambda \nabla \|x\|^2 &= 0 \\ \|x\|^2 &= 1 \end{aligned}$$

Now if $x = [x_j]$ then we have

$$\|Ax\|^2 = \sum_{k=1}^{n} \left(\sum_{j=1}^{n} a_{kj} x_j \right)^2$$

and for each i we see that

$$D_i \|Ax\|^2 = \sum_{k=1}^{n} 2 \left(\sum_{j=1}^{n} a_{kj} x_j \right) a_{ki} = 2 \sum_{k=1}^{n} \sum_{j=1}^{n} a_{kj} a_{ki} x_j.$$

Therefore the equation

$$\nabla \|Ax\|^2 - \lambda \nabla \|x\|^2 = 0$$

can be interpreted to say that

$$\sum_{k=1}^{n} \sum_{j=1}^{n} a_{kj} a_{ki} x_j - \lambda x_i = 0$$

for every $i = 1, 2, \cdots, n$.

If we write

$$\sum_{k=1}^{n} a_{ki} a_{kj} = b_{ij}$$

for all i and j then the condition

$$\sum_{k=1}^{n} \sum_{j=1}^{n} a_{kj} a_{ki} x_j - \lambda x_i = 0$$

can be written in the form

$$\sum_{j=1}^{n} b_{ij} x_j - \lambda x_i = 0.$$

We now interpret this condition in one more way. We define B to be the matrix $[b_{ij}]$. The preceding condition says that

$$(B - \lambda I)\, x = 0$$

where I is the identity matrix. Since $x \neq 0$ we deduce that the number λ must be an eigenvalue of the matrix B and that x must be an eigenvector corresponding to λ.

We can therefore use the following procedure for finding the norm of the matrix A:

1. Calculate the matrix B. This part of the procedure is easy because the definition

$$\sum_{k=1}^{n} a_{ki} a_{kj} = b_{ij}$$

tells us that $B = A^T A$ (where A^T stands for the transpose of the matrix A). Thus, we can obtain the matrix B by pointing at the expression $A^T A$, holding down the control key and clicking Evaluate.

2. Point at the matrix B and click on Matrices and Eigenvectors. You will see a formula for an eigenvector in terms of a variable ρ that is a root of a certain polynomial that we shall call $f(\rho)$. Don't worry if this eigenvector looks messy and doesn't fit onto your screen. Calling this vector v, for the moment, obtain a unit eigenvector by pointing at the equation

$$x = \frac{v}{\|v\|}$$

and clicking on Define and New Definition.

3. Point at the polynomial $f(\rho)$ and click on Polynomials and Roots, and you will see approximations to the eigenvalues of B.

4. For each eigenvalue ρ of the matrix B, supply *Scientific Notebook* with the definition of ρ and then point at the expression $\|Ax\|$ and click on Evaluate Numerically. The largest of these numbers is the norm of the given matrix A.
This last step may not work with some vintages of *Scientific Notebook*. If you have problems, look at the alternative approach suggested in Subsection 8.3.5.

8.3.4 Illustrating the Procedure

To illustrate the procedure just described for finding the norm of a matrix we shall take

$$A = \begin{bmatrix} -1 & -7 & -1 & 4 & -5 \\ -9 & 5 & -1 & 4 & 3 \\ -8 & 3 & -4 & 9 & -5 \\ 8 & -5 & 8 & 9 & 7 \\ 1 & 2 & -9 & -5 & -8 \end{bmatrix}.$$

Before we begin we supply this definition of the matrix A to *Scientific Notebook* by clicking on Define and New Definition. We now apply the method described in Subsection 8.3.3.

1. To find the matrix B we point at the expression $A^T A$ and click on Evaluate.

$$\begin{bmatrix} 211 & -100 & 97 & -45 & 66 \\ -100 & 112 & -68 & -36 & -16 \\ 97 & -68 & 163 & 73 & 150 \\ -45 & -36 & 73 & 219 & 50 \\ 66 & -16 & 150 & 50 & 172 \end{bmatrix}.$$

2. Now that we have found the matrix B we point at it and click on Matrices and Eigenvectors. We obtain[12]

$$\left\{ \begin{bmatrix} -\frac{556\,99745\,89639}{8804\,34201\,26647} - \frac{332\,89684\,65089}{5\,63477\,88881\,05408}\rho^2 - \frac{27219067}{22\,53911\,55524\,21632}\rho^4 + \frac{3\,57007\,39843}{22\,53911\,55524\,21632}\rho^3 + \frac{1855\,905}{35217\,368} \\ 1 \\ \frac{40437\,69028\,38233}{17608\,68402\,53294} + \frac{450\,92932\,56811}{2\,81738\,94440\,52704}\rho^2 + \frac{594542533}{90\,15646\,22096\,86528}\rho^4 - \frac{52\,81917\,53649}{90\,15646\,22096\,86528}\rho^3 - \frac{19961\,538}{1\,40869\,472} \\ \frac{7794\,26324\,43367}{1\,40869\,47220\,26352}\rho + \frac{49\,14659\,92105}{90\,15646\,22096\,86528}\rho^3 - \frac{694979117}{90\,15646\,22096\,86528}\rho^4 - \frac{581\,19208\,23677}{5\,63477\,88881\,05408}\rho^2 - \frac{4121\,89}{17608\,6} \\ \frac{12097\,45911\,63297}{1\,40869\,47220\,26352}\rho + \frac{24\,64979\,74697}{90\,15646\,22096\,86528}\rho^3 - \frac{282626077}{90\,15646\,22096\,86528}\rho^4 - \frac{222\,30589\,61407}{2\,81738\,94440\,52704}\rho^2 - \frac{32362\,6}{17608\,6} \end{bmatrix}$$

$\rho$where ρ is a root of $Z^5 - 877Z^4 + 241680Z^3 - 22992960Z^2 + 509746688Z -$

1124663296.

[12] The following expression will be visible only in the on-screen version of this text.

Change the symbol Z in this polynomial to ρ and call the polynomial $f(\rho)$

$$f(\rho) = \rho^5 - 877\rho^4 + 241680\rho^3 - 22992960\rho^2 + 509746688\rho - 11246\,63296.$$

Now point at the equation[13]

$$x = \frac{\begin{bmatrix} -\frac{556\,99745\,89639}{8804\,34201\,26647} - \frac{332\,89684\,65089}{5\,63477\,88881\,05408}\rho^2 - \frac{27219067}{22\,53911\,55524\,21632}\rho^4 + \frac{3\,57007\,39843}{22\,53911\,55524\,21632}\rho^3 + \\ 1 \\ \frac{40437\,69028\,38233}{17608\,68402\,53294} + \frac{450\,92932\,56811}{2\,81738\,94440\,52704}\rho^2 + \frac{594542533}{90\,15646\,22096\,86528}\rho^4 - \frac{52\,81917\,53649}{90\,15646\,22096\,86528}\rho^3 - \frac{1}{1} \\ \frac{7794\,26324\,43367}{1\,40869\,47220\,26352}\rho + \frac{49\,14659\,92105}{90\,15646\,22096\,86528}\rho^3 - \frac{694979117}{90\,15646\,22096\,86528}\rho^4 - \frac{581\,19208\,23677}{5\,63477\,88881\,05408}\rho^2 - \\ \frac{12097\,45911\,63297}{1\,40869\,47220\,26352}\rho + \frac{24\,64979\,74697}{90\,15646\,22096\,86528}\rho^3 - \frac{282626077}{90\,15646\,22096\,86528}\rho^4 - \frac{222\,30589\,61407}{2\,81738\,94440\,52704}\rho^2 - \end{bmatrix}}{\left\| \begin{bmatrix} -\frac{556\,99745\,89639}{8804\,34201\,26647} - \frac{332\,89684\,65089}{5\,63477\,88881\,05408}\rho^2 - \frac{27219067}{22\,53911\,55524\,21632}\rho^4 + \frac{3\,57007\,39843}{22\,53911\,55524\,21632}\rho^3 + \\ 1 \\ \frac{40437\,69028\,38233}{17608\,68402\,53294} + \frac{450\,92932\,56811}{2\,81738\,94440\,52704}\rho^2 + \frac{594542533}{90\,15646\,22096\,86528}\rho^4 - \frac{52\,81917\,53649}{90\,15646\,22096\,86528}\rho^3 - \frac{}{1} \\ \frac{7794\,26324\,43367}{1\,40869\,47220\,26352}\rho + \frac{49\,14659\,92105}{90\,15646\,22096\,86528}\rho^3 - \frac{694979117}{90\,15646\,22096\,86528}\rho^4 - \frac{581\,19208\,23677}{5\,63477\,88881\,05408}\rho^2 - \\ \frac{12097\,45911\,63297}{1\,40869\,47220\,26352}\rho + \frac{24\,64979\,74697}{90\,15646\,22096\,86528}\rho^3 - \frac{282626077}{90\,15646\,22096\,86528}\rho^4 - \frac{222\,30589\,61407}{2\,81738\,94440\,52704}\rho^2 - \end{bmatrix} \right\|}$$

and click on Define and New Definition. x

3. Point at the polynomial $f(\rho)$ and click on Polynomials and Roots. You will obtain

$$\begin{array}{c} 2.4756 \\ 27.663 \\ 140.65 \\ 264.08 \\ 442.12 \end{array}.$$

Highlight the equation $\rho = 2.4756$ and click on Define and New Definition. Then point to the symbol $\|Ax\|$ and click on Evaluate Numerically. You will obtain $1.57341\,6$. Repeat this procedure for each of the other four values of ρ to obtain the four other values of $\|Ax\|$. The five numbers come out as $1.57341\,6$, $5.25956\,9$, 11.85972, 16.25066 and 21.02676 and so the approximate value of $\|A\|$ is 21.02676. We note that this answer is the same as the one we obtained In Subsection 8.3.2.

8.3.5 In Case of Trouble

In some vintages of the *Scientific Notebook* link to Maple, the procedure described above for evaluating $\|Ax\|$ will not work properly. If you have problems, ignore the definition of the vector x that you have supplied to *Scientific Notebook* and work directly with the expression[14]

13 The following expression will be visible only in the on-screen version of this text.

14 The following expression will be visible only in the on-screen version of this text.

$$
A\frac{\left\|\left[\begin{matrix}
-\frac{556\,99745\,89639}{8804\,34201\,26647}-\frac{332\,89684\,65089}{5\,63477\,88881\,05408}\rho^2-\frac{27219067}{22\,53911\,55524\,21632}\rho^4+\frac{3\,57007\,39843}{22\,53911\,55524\,21632}\rho^3+\frac{1855\,9058}{35217\,368}\\
1\\
\frac{40437\,69028\,38233}{17608\,68402\,53294}+\frac{450\,92932\,56811}{2\,81738\,94440\,52704}\rho^2+\frac{594542533}{90\,15646\,22096\,86528}\rho^4-\frac{52\,81917\,53649}{90\,15646\,22096\,86528}\rho^3-\frac{19961\,5385}{1\,40869\,472}\\
\frac{7794\,26324\,43367}{1\,40869\,47220\,26352}\rho+\frac{49\,14659\,92105}{90\,15646\,22096\,86528}\rho^3-\frac{694979117}{90\,15646\,22096\,86528}\rho^4-\frac{581\,19208\,23677}{5\,63477\,88881\,05408}\rho^2-\frac{4121\,895}{17608\,68}\\
\frac{12097\,45911\,63297}{1\,40869\,47220\,26352}\rho+\frac{24\,64979\,74697}{90\,15646\,22096\,86528}\rho^3-\frac{282626077}{90\,15646\,22096\,86528}\rho^4-\frac{222\,30589\,61407}{2\,81738\,94440\,52704}\rho^2-\frac{32362\,66}{17608\,68}
\end{matrix}\right.\right\|}{\left\|\left[\begin{matrix}
-\frac{556\,99745\,89639}{8804\,34201\,26647}-\frac{332\,89684\,65089}{5\,63477\,88881\,05408}\rho^2-\frac{27219067}{22\,53911\,55524\,21632}\rho^4+\frac{3\,57007\,39843}{22\,53911\,55524\,21632}\rho^3+\frac{1855\,9058}{35217\,3680}\\
1\\
\frac{40437\,69028\,38233}{17608\,68402\,53294}+\frac{450\,92932\,56811}{2\,81738\,94440\,52704}\rho^2+\frac{594542533}{90\,15646\,22096\,86528}\rho^4-\frac{52\,81917\,53649}{90\,15646\,22096\,86528}\rho^3-\frac{19961\,5385}{1\,40869\,4722}\\
\frac{7794\,26324\,43367}{1\,40869\,47220\,26352}\rho+\frac{49\,14659\,92105}{90\,15646\,22096\,86528}\rho^3-\frac{694979117}{90\,15646\,22096\,86528}\rho^4-\frac{581\,19208\,23677}{5\,63477\,88881\,05408}\rho^2-\frac{4121\,895}{17608\,68}\\
\frac{12097\,45911\,63297}{1\,40869\,47220\,26352}\rho+\frac{24\,64979\,74697}{90\,15646\,22096\,86528}\rho^3-\frac{282626077}{90\,15646\,22096\,86528}\rho^4-\frac{222\,30589\,61407}{2\,81738\,94440\,52704}\rho^2-\frac{32362\,66}{17608\,68}
\end{matrix}\right.\right\|}
$$

by highlighting the fraction to the right of the symbol A, holding down the control key and clicking **twice** on Evaluate Numerically, and then finally, pointing to the entire expression and clicking on Evaluate Numerically.

8.4 Geometric Sequences of Matrices

A **geometric sequence** of matrices is a sequence of the form (A^n) where A is a square matrix. In this section we investigate the limits of sequences of this type.

8.4.1 Theorem: Convergence of Geometric Sequences

Suppose that A is a square matrix. Then the following two conditions are equivalent:

1. *There exists a matrix B such that $A^n \to B$ as $n \to \infty$.*
2. (a) *Given any (real or complex) eigenvalue λ of A, either $\lambda = 1$ or $|\lambda| < 1$.*
   (b) *In the event that the number 1 is an eigenvalue of A then the dimension of the space of eigenvectors corresponding to the eigenvalue 1 is equal to the multiplicity of the eigenvalue 1.*

Finally, if $A^n \to B$ as $n \to \infty$, then the matrix B is the zero matrix if and only if every eigenvalue λ of A satisfies $|\lambda| < 1$.

8.4.2 Convergence of Geometric Series

Suppose that A is a square matrix and that every eigenvalue λ of A satisfies $|\lambda| < 1$. From the identity

$$(I-A)\sum_{j=0}^{n} A^j = I - A^{n+1}$$

and the fact that the matrix $I - A$ has an inverse we deduce that

$$\sum_{j=0}^{n} A^j = (I-A)^{-1}\left(I - A^{n+1}\right) \to (I-A)^{-1}$$

as $n \to \infty$.

8.4.3 A Geometric Sequence that Converges to Zero

We define

$$A = \begin{bmatrix} -.45 & -.55 & -.37 & -.35 \\ .37 & .5 & -.39 & .5 \\ .21 & -.17 & .37 & -.59 \\ .45 & -.08 & -.53 & .52 \end{bmatrix}.$$

In order to work numerically with this matrix we have written its entries in decimal form. Point at the formula for A and click on Define and New Definition. Now point at the symbol A and click on Matrices and Eigenvalues. You will obtain the eigenvalues $-.22575 + .54238i$, $-.22575 - .54238i$, $.99466$ and $.39684$. Since the absolute value of each of these numbers is less than 1 we deduce that A^n approaches the zero matrix as $n \to \infty$.

Now we verify our conclusion by working out A^n for some large positive integers n.

$$A^{10} = \begin{bmatrix} -3.3913\times10^{-2} & 4.9272\times10^{-3} & .198 & -.21827 \\ .12756 & -2.1526\times10^{-2} & -.70193 & .75706 \\ -.10733 & 2.1704\times10^{-2} & .61093 & -.65383 \\ 6.9637\times10^{-2} & -9.5015\times10^{-3} & -.36758 & .39927 \end{bmatrix}$$

$$A^{20} = \begin{bmatrix} -3.4672\times10^{-2} & 6.0982\times10^{-3} & .19102 & -.20548 \\ .12098 & -2.1336\times10^{-2} & -.66674 & .71708 \\ -.10469 & 1.8476\times10^{-2} & .57708 & -.62064 \\ 6.3681\times10^{-2} & -1.1224\times10^{-2} & -.35087 & .37736 \end{bmatrix}$$

$$A^{30} = \begin{bmatrix} -3.2857\times10^{-2} & 5.7962\times10^{-3} & .18108 & -.19475 \\ .11467 & -2.0229\times10^{-2} & -.63198 & .67968 \\ -9.9249\times10^{-2} & 1.7509\times10^{-2} & .54699 & -.58828 \\ 6.0344\times10^{-2} & -1.0645\times10^{-2} & -.33258 & .35768 \end{bmatrix}$$

$$A^{100} = \begin{bmatrix} -2.2585\times10^{-2} & 3.9843\times10^{-3} & .12447 & -.13387 \\ 7.8822\times10^{-2} & -1.3905\times10^{-2} & -.43441 & .46721 \\ -6.8222\times10^{-2} & 1.2035\times10^{-2} & .37599 & -.40438 \\ .04148 & -7.3175\times10^{-3} & -.22861 & .24586 \end{bmatrix}.$$

So far, what we have seen is not very convincing. Now we look at

$$A^{500} = \begin{bmatrix} -2.6518 \times 10^{-3} & 4.6781 \times 10^{-4} & 1.4615 \times 10^{-2} & -1.5718 \times 10^{-2} \\ 9.2547 \times 10^{-3} & -1.6326 \times 10^{-3} & -5.1006 \times 10^{-2} & 5.4856 \times 10^{-2} \\ -8.0102 \times 10^{-3} & 1.4131 \times 10^{-3} & 4.4147 \times 10^{-2} & -4.7479 \times 10^{-2} \\ 4.8703 \times 10^{-3} & -8.5917 \times 10^{-4} & -2.6842 \times 10^{-2} & 2.8868 \times 10^{-2} \end{bmatrix}.$$

We begin to see that A^n is close to the zero matrix if n is large enough. Note finally that[15]

$$\begin{aligned} A^{5000} &= \left(A^{500}\right)^{10} \\ &= \begin{bmatrix} -9.0758 \times 10^{-14} & 1.6011 \times 10^{-14} & 5.002 \times 10^{-13} & -5.3795 \times 10^{-13} \\ 3.1675 \times 10^{-13} & -5.5878 \times 10^{-14} & -1.7457 \times 10^{-12} & 1.8775 \times 10^{-12} \\ -2.7415 \times 10^{-13} & 4.8363 \times 10^{-14} & 1.5109 \times 10^{-12} & -1.625 \times 10^{-12} \\ 1.6669 \times 10^{-13} & -2.9406 \times 10^{-14} & -9.1867 \times 10^{-13} & 9.8801 \times 10^{-13} \end{bmatrix} \end{aligned}$$

So, as long as we are patient enough, this numerical procedure does demonstrate that A^n approaches the zero matrix as $n \to \infty$.

8.4.4 A Convergent Sequence with a Nonzero Limit

In this example we take

$$A = \begin{bmatrix} -\frac{32}{441} & 0 & \frac{316}{49} & \frac{3149}{441} & -\frac{6014}{441} & \frac{3086}{441} \\ \frac{211}{441} & \frac{1}{3} & \frac{148}{147} & -\frac{19}{441} & -\frac{824}{441} & \frac{548}{441} \\ \frac{215}{441} & \frac{1}{3} & -\frac{884}{147} & -\frac{3098}{441} & \frac{6107}{441} & -\frac{3056}{441} \\ -\frac{263}{441} & 0 & \frac{120}{49} & \frac{1616}{441} & -\frac{2570}{441} & \frac{1406}{441} \\ -\frac{2}{147} & \frac{1}{3} & -\frac{293}{147} & -\frac{277}{147} & \frac{682}{147} & -\frac{284}{147} \\ \frac{4}{49} & \frac{1}{3} & \frac{1}{147} & \frac{1}{49} & \frac{59}{147} & \frac{15}{49} \end{bmatrix}$$

and we supply this definition to *Scientific Notebook* by clicking on Define and New Definition. We now explore the sequence (A^n) by working out A^n numerically for some large positive integers n.

$$A^{20} = \begin{bmatrix} -.28571 & -.14286 & -.14284 & .28573 & -.14288 & 1.3162 \times 10^{-5} \\ .57143 & .28571 & .28572 & -.57143 & .28571 & 1.9131 \times 10^{-6} \\ .85714 & .42857 & .42856 & -.85716 & .4286 & -1.3568 \times 10^{-5} \\ -.28571 & -.14286 & -.14285 & .28572 & -.14287 & 5.0916 \times 10^{-6} \\ .57143 & .28571 & .28571 & -.57143 & .28572 & -4.4851 \times 10^{-6} \\ .57143 & .28571 & .28571 & -.57143 & .28572 & -4.4972 \times 10^{-7} \end{bmatrix}$$

15 The following expression will be fully visible only in the on-screen version of this text.

$$A^{50} = \begin{bmatrix} -.28571 & -.14286 & -.14286 & .28571 & -.14286 & 3.0019 \times 10^{-16} \\ .57143 & .28571 & .28571 & -.57143 & .28571 & 4.8402 \times 10^{-17} \\ .85714 & .42857 & .42857 & -.85714 & .42857 & -3.1321 \times 10^{-16} \\ -.28571 & -.14286 & -.14286 & .28571 & -.14286 & 1.1471 \times 10^{-16} \\ .57143 & .28571 & .28571 & -.57143 & .28571 & -1.0616 \times 10^{-16} \\ .57143 & .28571 & .28571 & -.57143 & .28571 & -1.3415 \times 10^{-17} \end{bmatrix}$$

$$A^{100} = \begin{bmatrix} -.28572 & -.14286 & -.14286 & .28572 & -.14286 & 2.9977 \times 10^{-21} \\ .57142 & .28571 & .28571 & -.57142 & .28571 & -4.309 \times 10^{-22} \\ .85714 & .42857 & .42857 & -.85714 & .42857 & -3.4286 \times 10^{-21} \\ -.28572 & -.14286 & -.14286 & .28572 & -.14286 & 2.9977 \times 10^{-21} \\ .57142 & .28571 & .28571 & -.57142 & .28571 & -4.309 \times 10^{-22} \\ .57142 & .28571 & .28571 & -.57142 & .28571 & -4.309 \times 10^{-22} \end{bmatrix}$$

$$A^{200} = \begin{bmatrix} -.28572 & -.14286 & -.14286 & .28572 & -.14286 & 6.1293 \times 10^{-22} \\ .57141 & .28571 & .28571 & -.57141 & .28571 & -1.2258 \times 10^{-21} \\ .85713 & .42857 & .42857 & -.85713 & .42857 & -1.8387 \times 10^{-21} \\ -.28572 & -.14286 & -.14286 & .28572 & -.14286 & 6.1293 \times 10^{-22} \\ .57141 & .28571 & .28571 & -.57141 & .28571 & -1.2258 \times 10^{-21} \\ .57141 & .28571 & .28571 & -.57141 & .28571 & -1.2258 \times 10^{-21} \end{bmatrix}$$

$$A^{4000} = \begin{bmatrix} -.28566 & -.14283 & -.14283 & .28566 & -.14283 & 6.128 \times 10^{-22} \\ .5713 & .28566 & .28566 & -.5713 & .28566 & -1.2256 \times 10^{-21} \\ .85696 & .42849 & .42849 & -.85696 & .42849 & -1.8384 \times 10^{-21} \\ -.28566 & -.14283 & -.14283 & .28566 & -.14283 & 6.128 \times 10^{-22} \\ .5713 & .28566 & .28566 & -.5713 & .28566 & -1.2256 \times 10^{-21} \\ .5713 & .28566 & .28566 & -.5713 & .28566 & -1.2256 \times 10^{-21} \end{bmatrix}.$$

It seems clear that the sequence (A^n) is convergent to a matrix whose last column is zero. Now we look at this convergence in view of Theorem 8.4.1.When we ask *Scientific Notebook* for the eigenvalues of A we obtain the numbers A, eigenvalues: $1, \frac{3}{7}, \frac{3}{7}, \frac{1}{3}, \frac{1}{3}, \frac{1}{3}$. Since the eigenvalue 1 has a multiplicity of only 1 and the other eigenvalues have absolute values less than 1, Theorem 8.4.1 confirms that the sequence is convergent.

Finally, we remark that the limit of the sequence (A^n) can be found by expressing A in terms of a matrix of Jordan blocks. It is not hard to show that

$$A = P^{-1} \begin{bmatrix} \frac{1}{3} & 1 & -2 & 0 & 0 & 0 \\ 0 & \frac{1}{3} & -1 & 0 & 0 & 0 \\ 0 & 0 & \frac{1}{3} & 0 & 0 & 0 \\ 0 & 0 & 0 & 1 & 0 & 0 \\ 0 & 0 & 0 & 0 & \frac{3}{7} & 2 \\ 0 & 0 & 0 & 0 & 0 & \frac{3}{7} \end{bmatrix} P$$

where

$$P = \begin{bmatrix} 1 & -2 & 1 & 2 & 1 & 1 \\ 0 & 1 & 1 & 3 & 1 & -2 \\ 1 & 0 & 0 & -1 & 2 & -2 \\ 2 & 1 & 1 & -2 & 1 & 0 \\ 1 & 0 & 0 & -1 & 1 & -1 \\ 0 & 0 & 1 & 1 & -2 & 1 \end{bmatrix}$$

and so, if n is a positive integer then we have

$$A^n = P^{-1} \begin{bmatrix} \frac{1}{3} & 1 & -2 & 0 & 0 & 0 \\ 0 & \frac{1}{3} & -1 & 0 & 0 & 0 \\ 0 & 0 & \frac{1}{3} & 0 & 0 & 0 \\ 0 & 0 & 0 & 1 & 0 & 0 \\ 0 & 0 & 0 & 0 & \frac{3}{7} & 2 \\ 0 & 0 & 0 & 0 & 0 & \frac{3}{7} \end{bmatrix}^n P.$$

It is easy to see that

$$\begin{bmatrix} \frac{1}{3} & 1 & -2 & 0 & 0 & 0 \\ 0 & \frac{1}{3} & -1 & 0 & 0 & 0 \\ 0 & 0 & \frac{1}{3} & 0 & 0 & 0 \\ 0 & 0 & 0 & 1 & 0 & 0 \\ 0 & 0 & 0 & 0 & \frac{3}{7} & 2 \\ 0 & 0 & 0 & 0 & 0 & \frac{3}{7} \end{bmatrix}^n \rightarrow \begin{bmatrix} 0 & 0 & 0 & 0 & 0 & 0 \\ 0 & 0 & 0 & 0 & 0 & 0 \\ 0 & 0 & 0 & 0 & 0 & 0 \\ 0 & 0 & 0 & 1 & 0 & 0 \\ 0 & 0 & 0 & 0 & 0 & 0 \\ 0 & 0 & 0 & 0 & 0 & 0 \end{bmatrix}$$

as $n \rightarrow \infty$ and therefore

$$\lim_{n\rightarrow\infty} A^n = P^{-1} \begin{bmatrix} 0 & 0 & 0 & 0 & 0 & 0 \\ 0 & 0 & 0 & 0 & 0 & 0 \\ 0 & 0 & 0 & 0 & 0 & 0 \\ 0 & 0 & 0 & 1 & 0 & 0 \\ 0 & 0 & 0 & 0 & 0 & 0 \\ 0 & 0 & 0 & 0 & 0 & 0 \end{bmatrix} P = \begin{bmatrix} -\frac{2}{7} & -\frac{1}{7} & -\frac{1}{7} & \frac{2}{7} & -\frac{1}{7} & 0 \\ \frac{4}{7} & \frac{2}{7} & \frac{2}{7} & -\frac{4}{7} & \frac{2}{7} & 0 \\ \frac{6}{7} & \frac{3}{7} & \frac{3}{7} & -\frac{6}{7} & \frac{3}{7} & 0 \\ -\frac{2}{7} & -\frac{1}{7} & -\frac{1}{7} & \frac{2}{7} & -\frac{1}{7} & 0 \\ \frac{4}{7} & \frac{2}{7} & \frac{2}{7} & -\frac{4}{7} & \frac{2}{7} & 0 \\ \frac{4}{7} & \frac{2}{7} & \frac{2}{7} & -\frac{4}{7} & \frac{2}{7} & 0 \end{bmatrix}$$

This last remark also provides a hint at the way in which Theorem 8.4.1 may be proved.

8.4.5 A Divergent Geometric Sequence

In this example we take

$$A = \begin{bmatrix} -\frac{18}{35} & \frac{2}{7} & \frac{22}{49} & \frac{551}{245} & -\frac{1108}{735} & \frac{306}{245} \\ -\frac{6}{35} & \frac{3}{7} & -\frac{16}{49} & -\frac{178}{245} & -\frac{241}{735} & \frac{67}{245} \\ \frac{113}{105} & -\frac{4}{21} & \frac{47}{147} & -\frac{941}{735} & \frac{1511}{735} & -\frac{477}{245} \\ -\frac{11}{35} & \frac{2}{7} & \frac{22}{49} & \frac{502}{245} & -\frac{226}{735} & \frac{12}{245} \\ \frac{17}{105} & \frac{2}{21} & \frac{50}{147} & \frac{691}{735} & \frac{739}{735} & -\frac{178}{245} \\ -\frac{46}{105} & \frac{2}{21} & \frac{50}{147} & \frac{1132}{735} & -\frac{437}{735} & \frac{214}{245} \end{bmatrix}$$

and we supply *Scientific Notebook* with the definition of this matrix by clicking on Define and New Definition. We now explore the sequence (A^n) by working out A^n numerically for some large positive integers n.[16]

$$A^{20} = \begin{bmatrix} 39.0 & 3.9524 & 4.2381 & -26.286 & 82.048 & -86.0 \\ -63.0 & -6.9048 & -7.4762 & 40.571 & -133.1 & 140.0 \\ 13.0 & 1.4762 & 1.619 & -8.1429 & 27.524 & -29.0 \\ 40.0 & 3.9524 & 4.2381 & -27.286 & 84.048 & -88.0 \\ 52.0 & 5.4286 & 5.8571 & -34.429 & 109.57 & -115.0 \\ 51.0 & 5.4286 & 5.8571 & -33.429 & 107.57 & -113.0 \end{bmatrix}$$

$$A^{50} = \begin{bmatrix} 243.29 & 9.6667 & 9.9524 & -213.43 & 496.33 & -506.0 \\ -441.57 & -18.333 & -18.905 & 384.86 & -901.67 & 920.0 \\ 100.14 & 4.3333 & 4.4762 & -86.714 & 204.67 & -209.0 \\ 244.29 & 9.6667 & 9.9524 & -214.43 & 498.33 & -508.0 \\ 343.43 & 14.0 & 14.429 & -300.14 & 701.0 & -715.0 \\ 342.43 & 14.0 & 14.429 & -299.14 & 699.0 & -713.0 \end{bmatrix}$$

$$A^{100} = \begin{bmatrix} 964.35 & 19.175 & 19.45 & -905.92 & 1947.7 & -1966.9 \\ -1834.9 & -37.401 & -37.954 & 1720.9 & -3706.8 & 3744.4 \\ 436.44 & 9.1194 & 9.2587 & -408.63 & 881.91 & -891.07 \\ 965.35 & 19.175 & 19.45 & -906.92 & 1949.7 & -1968.9 \\ 1401.8 & 28.335 & 28.75 & -1315.4 & 2831.6 & -2860.1 \\ 1400.8 & 28.335 & 28.75 & -1314.4 & 2829.6 & -2858.1 \end{bmatrix}$$

16 One of the expressions that follow will be fully visible only in the on-screen version of this text.

$$A^{200} = \begin{bmatrix} 3797.9 & 36.571 & 36.777 & -3687.8 & 7634.6 & -7666.4 \\ -7189.3 & -67.837 & -68.187 & 6985.1 & -14451. & 14509. \\ 1768.2 & 17.067 & 17.16 & -1716.8 & 3554.5 & -3569.3 \\ 3798.9 & 36.571 & 36.777 & -3688.8 & 7636.6 & -7668.4 \\ 5469.0 & 51.598 & 51.866 & -5313.7 & 10993. & -11037. \\ 5468.0 & 51.598 & 51.866 & -5312.7 & 10991. & -11035. \end{bmatrix}$$

$$A^{1000} = \begin{bmatrix} -1.8076\times 10^{7} & -1.2939\times 10^{5} & -1.289\times 10^{5} & 1.7688\times 10^{7} & -3.6305\times 10^{7} & 3. \\ 3.129\times 10^{7} & 2.0683\times 10^{5} & 2.0537\times 10^{5} & -3.0672\times 10^{7} & 6.2834\times 10^{7} & -6 \\ -8.3468\times 10^{6} & -58931. & -58677. & 8.1703\times 10^{6} & -1.6764\times 10^{7} & 1. \\ -1.809\times 10^{7} & -1.2953\times 10^{5} & -1.2904\times 10^{5} & 1.7703\times 10^{7} & -3.6334\times 10^{7} & 3. \\ -2.3533\times 10^{7} & -1.5345\times 10^{5} & -1.5228\times 10^{5} & 2.3074\times 10^{7} & -4.7255\times 10^{7} & 4. \\ -2.3529\times 10^{7} & -1.5341\times 10^{5} & -1.5224\times 10^{5} & 2.307\times 10^{7} & -4.7247\times 10^{7} & 4. \end{bmatrix}$$

It seems clear that the sequence (A^n) is not approaching any finite limit. Now we look at this divergence in view of Theorem 8.4.1.When we ask *Scientific Notebook* for the eigenvalues of A we obtain the numbers $\frac{3}{7}$, $\frac{1}{3}$, $\frac{2}{5}$, 1, 1 and 1 and we can see that the eigenvalue 1 has multiplicity three. However, when we point the matrix A again and click on Eigenvectors we obtain

$$\left\{\begin{bmatrix} 1 \\ \frac{3}{2} \\ -3 \\ 1 \\ \frac{3}{2} \\ \frac{3}{2} \end{bmatrix}\right\} \leftrightarrow \frac{3}{7}, \left\{\begin{bmatrix} 1 \\ -2 \\ \frac{1}{2} \\ 1 \\ \frac{3}{2} \\ \frac{3}{2} \end{bmatrix}\right\} \leftrightarrow 1, \left\{\begin{bmatrix} 1 \\ -2 \\ -3 \\ 1 \\ -2 \\ -2 \end{bmatrix}\right\} \leftrightarrow \frac{1}{3}, \left\{\begin{bmatrix} \frac{34}{13} \\ \frac{11}{26} \\ -\frac{71}{26} \\ 1 \\ -\frac{12}{13} \\ -\frac{3}{26} \end{bmatrix}\right\} \leftrightarrow \frac{2}{5}$$

which shows that the space of eigenvectors corresponding to the eigenvalue 1 has dimension 1 only. Thus Theorem 8.4.1 confirms that the sequence (A^n) is divergent.

8.4.6 The Rabbits and Foxes Problem

In this example we are concerned with the populations of rabbits and foxes in a national park. Suppose that 1000 rabbits and 1000 foxes are introduced into the park which previously contained no rabbits or foxes and for every nonnegative integer n, the numbers of rabbits and foxes in the path after n months are $R(n)$ and $F(n)$, respectively. Suppose finally that for each n we have

$$\begin{aligned} R(n+1) &= 1.1R(n) - 0.2F(n) \\ F(n+1) &= 0.2R(n) + 0.6F(n) \end{aligned}$$

We want to determine what will happen to the populations of rabbits and foxes in the long term.

We begin our study of this problem with the observation that if

$$A = \begin{bmatrix} 1.1 & -0.2 \\ 0.2 & 0.6 \end{bmatrix}$$

then, for each n,

$$\begin{bmatrix} R(n+1) \\ F(n+1) \end{bmatrix} = \begin{bmatrix} 1.1 & -0.2 \\ 0.2 & 0.6 \end{bmatrix} \begin{bmatrix} R(n) \\ F(n) \end{bmatrix}$$

and so

$$\begin{bmatrix} R(n) \\ F(n) \end{bmatrix} = \begin{bmatrix} 1.1 & -0.2 \\ 0.2 & 0.6 \end{bmatrix}^n \begin{bmatrix} 1000 \\ 1000 \end{bmatrix}.$$

By pointing at the matrix A and clicking on Eigenvalues we see that the two eigenvalues of this matrix are 1 and 0.7. Since the eigenvalue 1 has multiplicity only 1 and the other eigenvalue has absolute value less than 1, Theorem 8.4.1 tells us that the sequence (A^n) is convergent.

8.4.6.1 Solving the Problem Numerically

By clicking on Evaluate Numerically we see that

$$\begin{bmatrix} 1.1 & -0.2 \\ 0.2 & 0.6 \end{bmatrix}^5 = \begin{bmatrix} 1.2773 & -.55462 \\ .55462 & -.10924 \end{bmatrix}$$

$$\begin{bmatrix} 1.1 & -0.2 \\ 0.2 & 0.6 \end{bmatrix}^{20} = \begin{bmatrix} 1.3331 & -.66613 \\ .66613 & -.33227 \end{bmatrix}$$

$$\begin{bmatrix} 1.1 & -0.2 \\ 0.2 & 0.6 \end{bmatrix}^{40} = \begin{bmatrix} 1.3333 & -.66667 \\ .66667 & -.33333 \end{bmatrix}$$

$$\begin{bmatrix} 1.1 & -0.2 \\ 0.2 & 0.6 \end{bmatrix}^{200} = \begin{bmatrix} 1.3333 & -.66667 \\ .66667 & -.33333 \end{bmatrix}.$$

Looking at these matrices we can conjecture that

$$\begin{bmatrix} 1.1 & -0.2 \\ 0.2 & 0.6 \end{bmatrix}^n \to \begin{bmatrix} \frac{4}{3} & -\frac{2}{3} \\ \frac{2}{3} & -\frac{1}{3} \end{bmatrix}$$

as $n \to \infty$ and that, in the long term, the numbers of rabbits and foxes will approach the coordinates of the vector

$$\begin{bmatrix} \frac{4}{3} & -\frac{2}{3} \\ \frac{2}{3} & -\frac{1}{3} \end{bmatrix} \begin{bmatrix} 1000 \\ 1000 \end{bmatrix} = \begin{bmatrix} 666.67 \\ 333.33 \end{bmatrix}.$$

In other words, in the long term, there will be twice as many rabbits in the park as there are foxes.

8.4.6.2 Solving the Problem Exactly

In order to work exactly with the matrix A we must rewrite it in a form that does not contain decimals. We therefore write the matrix A as

$$A = \begin{bmatrix} \frac{11}{10} & -\frac{1}{5} \\ \frac{1}{5} & \frac{3}{5} \end{bmatrix}$$

By pointing at the matrix A and clicking on Eigenvectors we obtain A,

$$\left\{\begin{bmatrix} 2 \\ 1 \end{bmatrix}\right\} \leftrightarrow 1, \left\{\begin{bmatrix} 1 \\ 2 \end{bmatrix}\right\} \leftrightarrow \frac{7}{10}$$

We define

$$P = \begin{bmatrix} 2 & 1 \\ 1 & 2 \end{bmatrix}$$

and we supply this definition to *Scientific Notebook* by clicking on Define and New Definition. Since

$$P^{-1}AP = \begin{bmatrix} 1 & 0 \\ 0 & \frac{7}{10} \end{bmatrix}$$

we deduce that if n is any positive integer then

$$A^n = P \begin{bmatrix} 1 & 0 \\ 0 & \frac{7}{10} \end{bmatrix}^n P^{-1}.$$

Thus

$$\lim_{n \to \infty} A^n = P \begin{bmatrix} 1 & 0 \\ 0 & 0 \end{bmatrix} P^{-1} = \begin{bmatrix} \frac{4}{3} & -\frac{2}{3} \\ \frac{2}{3} & -\frac{1}{3} \end{bmatrix}$$

showing that the conjecture we made previously was correct.

8.5 Markov Processes

In this section we give a brief and elementary introduction to the concept of Markov processes in finite probability spaces and we suggest some ways in which the computing features of *Scientific Notebook* can be used to draw conclusions about these Markov processes.

We begin with a simple application of Markov processes that will help to motivate the theory.

8.5.1 The Car Rental Problem

Let us suppose that a car rental agency has three locations: 1, 2 and 3 and that a customer can rent a car at any of the three locations and return it to any of the locations. For all i and j in $\{1,2,3\}$ we shall use the symbol p_{ij} to describe the probability that a customer who has rented a car at location j will return it to location i. We observe that if $j \in \{1,2,3\}$ then a car rented at location j must be returned to one of the three locations. Therefore

$$p_{1j} + p_{2j} + p_{3j} = 1.$$

Thus, if we define a matrix P by the equation

$$P = \begin{bmatrix} p_{11} & p_{12} & p_{13} \\ p_{21} & p_{22} & p_{23} \\ p_{31} & p_{32} & p_{33} \end{bmatrix}$$

then the sum of the entries in each column of the matrix P must be 1. This process of observing a car as it moves from location to location as it is repeatedly rented is known as a **Markov process** and the matrix P is called the **transition matrix** of the Markov process. The first column

$$\begin{bmatrix} p_{11} \\ p_{21} \\ p_{31} \end{bmatrix} = \begin{bmatrix} p_{11} & p_{12} & p_{13} \\ p_{21} & p_{22} & p_{23} \\ p_{31} & p_{32} & p_{33} \end{bmatrix} \begin{bmatrix} 1 \\ 0 \\ 0 \end{bmatrix}$$

of the matrix P lists the probabilities that a car that was originally at location 1 will be at the locations 1, 2 and 3 after it has been rented once. Similarly, the second and third columns

$$\begin{bmatrix} p_{12} \\ p_{22} \\ p_{32} \end{bmatrix} = \begin{bmatrix} p_{11} & p_{12} & p_{13} \\ p_{21} & p_{22} & p_{23} \\ p_{31} & p_{32} & p_{33} \end{bmatrix} \begin{bmatrix} 0 \\ 1 \\ 0 \end{bmatrix}$$

$$\begin{bmatrix} p_{13} \\ p_{23} \\ p_{33} \end{bmatrix} = \begin{bmatrix} p_{11} & p_{12} & p_{13} \\ p_{21} & p_{22} & p_{23} \\ p_{31} & p_{32} & p_{33} \end{bmatrix} \begin{bmatrix} 0 \\ 0 \\ 1 \end{bmatrix}$$

of the matrix P list the probabilities that a car that was originally at location 2 or 3 will be at the locations 1, 2 and 3 after it has been rented once.

Now suppose that a car could have originated at any of the three locations and that the probabilities that the car originated in the locations 1, 2 and 3 are written as x_1, x_2 and x_3, respectively. Note that x_1, x_2 and x_3 are nonnegative numbers and that $x_1 + x_2 + x_3 = 1$. The probabilities that this car will be at the locations 1, 2 and 3 after it has been rented once are the coordinates of the vector

$$\begin{bmatrix} p_{11} & p_{12} & p_{13} \\ p_{21} & p_{22} & p_{23} \\ p_{31} & p_{32} & p_{33} \end{bmatrix} \left(x_1 \begin{bmatrix} 1 \\ 0 \\ 0 \end{bmatrix} + x_2 \begin{bmatrix} 0 \\ 1 \\ 0 \end{bmatrix} + x_3 \begin{bmatrix} 0 \\ 0 \\ 1 \end{bmatrix} \right) = P \begin{bmatrix} x_1 \\ x_2 \\ x_3 \end{bmatrix}.$$

Repeating the process we see that if the vector

$$\mathbf{x} = \begin{bmatrix} x_1 \\ x_2 \\ x_3 \end{bmatrix}$$

lists the probabilities that a car originated at the locations 1, 2 and 3, then the probabilities that this car will be at the locations 1, 2 and 3 after it has been rented twice are the coordinates of the vector $P^2\mathbf{x}$. In general, if n is any nonnegative integer, the vector $P^n\mathbf{x}$ lists the probabilities that the car will be at the locations 1, 2 and 3 respectively after it has been rented n times.

We shall now consider the case in which the matrix P is given by the equation

$$P = \begin{bmatrix} 0.8 & 0.3 & 0.2 \\ 0.1 & 0.2 & 0.6 \\ 0.1 & 0.5 & 0.2 \end{bmatrix}.$$

Point at this equation and click on Define and New Definition in order to supply this definition of P to *Scientific Notebook.*

8.5.1.1 Numerical Approach to the Problem

We observe that

$$P^4 = \begin{bmatrix} .5971 & .5114 & .503 \\ .2086 & .2607 & .2506 \\ .1943 & .2279 & .2464 \end{bmatrix}$$

$$P^{12} = \begin{bmatrix} .5577 & .55701 & .55693 \\ .22934 & .2297 & .22974 \\ .21296 & .21328 & .21333 \end{bmatrix}$$

$$P^{20} = \begin{bmatrix} .55738 & .55737 & .55737 \\ .22951 & .22951 & .22951 \\ .21311 & .21312 & .21312 \end{bmatrix}$$

$$P^{100} = \begin{bmatrix} .55738 & .55738 & .55738 \\ .22951 & .22951 & .22951 \\ .21311 & .21311 & .21311 \end{bmatrix}.$$

Work out the matrix P^n for some other positive integers n. It seems clear that the sequence

(P^n) approaches a limit matrix that is approximately

$$\begin{bmatrix} .55738 & .55738 & .55738 \\ .22951 & .22951 & .22951 \\ .21311 & .21311 & .21311 \end{bmatrix}$$

and a particularly interesting feature of this limit matrix is that all of its columns are the same. In other words, if n is a sufficiently large positive integer then the probabilities that a car that originated at any of the three locations will be at the locations 1, 2 and 3, respectively after it has been rented n times are

$$\begin{bmatrix} .55738 \\ .22951 \\ .21311 \end{bmatrix}.$$

After the company has been in business for a long, about 55.7% of its cars will be at location 1, about 22.9% of its cars will be at location 2 and about 21.3% of its cars will be at location 2.

8.5.1.2 Exact Evaluation of this Limit

In order to work exactly with the matrix P we must write it in a form that does not involve any decimals. We write P in the form

$$P = \begin{bmatrix} \frac{4}{5} & \frac{3}{10} & \frac{1}{5} \\ \frac{1}{10} & \frac{1}{5} & \frac{3}{5} \\ \frac{1}{10} & \frac{1}{2} & \frac{1}{5} \end{bmatrix}$$

and we supply its definition to *Scientific Notebook*. By pointing at the matrix P and clicking on Eigenvectors we see that one eigenvector of P is

$$\begin{bmatrix} \frac{34}{13} \\ \frac{14}{13} \\ 1 \end{bmatrix}$$

and that the other two are given in the form

$$\begin{bmatrix} -\frac{5}{2}\rho - \frac{3}{4} \\ -\frac{1}{4} + \frac{5}{2}\rho \\ 1 \end{bmatrix}$$

where ρ is a root of the equation $100\rho^2 - 20\rho - 19 = 0$. Solving this equation and substituting its solutions in the preceding formula we see that the vectors

$$\begin{bmatrix} \frac{34}{13} \\ \frac{14}{13} \\ 1 \end{bmatrix}, \quad \begin{bmatrix} -1-\frac{1}{2}\sqrt{5} \\ \frac{1}{2}\sqrt{5} \\ 1 \end{bmatrix}, \quad \begin{bmatrix} -1+\frac{1}{2}\sqrt{5} \\ -\frac{1}{2}\sqrt{5} \\ 1 \end{bmatrix}$$

are all eigenvectors of P. We now define

$$U = \begin{bmatrix} \frac{34}{13} & -1-\frac{1}{2}\sqrt{5} & -1+\frac{1}{2}\sqrt{5} \\ \frac{14}{13} & \frac{1}{2}\sqrt{5} & -\frac{1}{2}\sqrt{5} \\ 1 & 1 & 1 \end{bmatrix}$$

and, by pointing at the matrix $U^{-1}PU$ and clicking on Evaluate, we see that

$$U^{-1}PU = \begin{bmatrix} 1 & 0 & 0 \\ 0 & \frac{1}{10}+\frac{1}{5}\sqrt{5} & 0 \\ 0 & 0 & \frac{1}{10}-\frac{1}{5}\sqrt{5} \end{bmatrix}.$$

Thus, if n is a positive integer we have

$$P^n = U \begin{bmatrix} 1 & 0 & 0 \\ 0 & \frac{1}{10}+\frac{1}{5}\sqrt{5} & 0 \\ 0 & 0 & \frac{1}{10}-\frac{1}{5}\sqrt{5} \end{bmatrix}^n U^{-1}$$

and so

$$\lim_{n\to\infty} P^n = U \begin{bmatrix} 1 & 0 & 0 \\ 0 & 0 & 0 \\ 0 & 0 & 0 \end{bmatrix} U^{-1} = \begin{bmatrix} \frac{34}{61} & \frac{34}{61} & \frac{34}{61} \\ \frac{14}{61} & \frac{14}{61} & \frac{14}{61} \\ \frac{13}{61} & \frac{13}{61} & \frac{13}{61} \end{bmatrix}.$$

8.5.2 Introduction to Markov Processes

Suppose that we have a "system" that could be in any one of k possible states (just as a car could be at any one of three locations in the preceding example).[17]

---

[17] More precisely, we suppose that we have a finite probability space with k points (events).

The system begins in any one of the k possible states (just as a car could originate in any one of the locations in the preceding example). We then perform a sequence of observations and for each of these we note the state in which the system presently exists (just as, in the preceding example, we observed the location of a car each time it was rented and returned by a customer). We assume that for each of these observations, the probability that the system will be seen to be in any given state j depends only upon the state in which the system had been prior to that observation.

For each $i \in \{1, 2, \cdots, k\}$ we write p_{ij} for the probability that the system which was in state j prior to a given observation will be found by that observation to be in state i. The number p_{ij} is called the **transition probability** that the system will move from state j to state i. We Define P to be the $k \times k$ matrix $[p_{ij}]$ and we call P the **transition matrix** of the Markov process. Note that for all i and j we have $0 \leq p_{ij} \leq 1$.

For each j, column j of the matrix P is

$$\begin{bmatrix} p_{1j} \\ p_{2j} \\ \vdots \\ p_{kj} \end{bmatrix} = Pe_j$$

where e_j is the jth standard basis vector of $\mathbf{R}^k$. This column lists the probabilities that the system, which was in state j just prior to a given observation, will be found to be in the states $1, 2, \cdots, k$ in that observation. Since the observation must find the system in one of these k possible states we have

$$\sum_{i=1}^{k} p_{ij} = 1.$$

Inn other words, the sum of the numbers in each column of the matrix P must be equal to 1. We call the columns of P, the **state vectors** of the system.

Suppose now that, at a given instant, the system could be in any of the k states and that the probability that it is in state j for any given j is written as x_j. Of course, $0 \leq x_j \leq 1$ for each j and $\sum_{j=1}^{k} x_j = 1$. The vector

$$\mathbf{x} = \begin{bmatrix} x_1 \\ x_2 \\ \vdots \\ x_k \end{bmatrix}$$

is said to be a **probability vector.** The probabilities that the system will be found in the next observation to be in the states $1, 2, \cdots, k$ are the coordinates of the vector

$$x_1 Pe_1 + \cdots + x_k Pe_k = P(x_1 e_1 + \cdots + x_k e_k) = P\mathbf{x}.$$

Repeating the process we see that if the vector $\mathbf{x}$ lists the probabilities that the system originated in the states $1, 2, \cdots, k$, then the probabilities that the system will be in these states after two observations will be

$$P(P\mathbf{x}) = P^2\mathbf{x}.$$

In general, if n is any nonnegative integer, the vector $P^n\mathbf{x}$ lists the probabilities that the system will be in the states $1, 2, \cdots, k$ after n observations. In particular, for each j, column j of the matrix P^n, which is also $P^n e_j$, lists the probabilities that the system will be in the states $1, 2, \cdots, k$ after n observations if it originated in state j. Column j of the matrix P^n is sometimes known as the state vector of the system at the nth observation.

8.5.3 Limiting Behavior

We now discuss the behavior of the matrices P^n where P is a transition matrix that satisfies an additional requirement known as **regularity.** We say that a transition matrix $P = [p_{ij}]$ is **regular** if the numbers p_{ij} are all positive.

8.5.3.1 Theorem: Existence of the Limit Matrix

Suppose that P is a regular transition matrix of order $k \times k$. Then there exists a probability vector

$$\mathbf{q} = \begin{bmatrix} q_1 \\ q_2 \\ \vdots \\ q_k \end{bmatrix}$$

such that

$$\lim_{n\to\infty} P^n \to \begin{bmatrix} q_1 & q_1 & \cdots & q_1 \\ q_2 & q_2 & \cdots & q_2 \\ \vdots & \vdots & \ddots & \vdots \\ q_k & q_k & \cdots & q_k \end{bmatrix}.$$

8.5.3.2 Behavior of the Limit Matrix

Suppose that P is a regular transition matrix and that

$$Q = \lim_{n\to\infty} P^n.$$

Since every column of the matrix Q is the vector

$$\mathbf{q} = \begin{bmatrix} q_1 \\ q_2 \\ \vdots \\ q_k \end{bmatrix},$$

the coordinates of $\mathbf{q}$ list the probabilities that the system will be in the states $1, 2, \cdots, k$ after a large number of observations, regardless of the state in which the system originated. Letting

$n \to \infty$ in the equation

$$PP^n = P^{n+1}$$

we deduce that

$$PQ = Q$$

and, in particular, $P\mathbf{q} = \mathbf{q}$, so that the vector $\mathbf{q}$ is an eigenvector of P and also of Q corresponding to the eigenvalue 1. Given any vector

$$\mathbf{x} = \begin{bmatrix} x_1 \\ x_2 \\ \vdots \\ x_k \end{bmatrix},$$

the equation $Q\mathbf{x} = 0$ says that

$$\begin{bmatrix} q_1 \left(\sum_{j=1}^{k} x_j\right) \\ q_2 \left(\sum_{j=1}^{k} x_j\right) \\ \vdots \\ q_k \left(\sum_{j=1}^{k} x_j\right) \end{bmatrix} = \begin{bmatrix} 0 \\ 0 \\ \vdots \\ 0 \end{bmatrix},$$

in other words, $\sum_{j=1}^{k} x_j = 0$. We deduce that the matrix has an eigenvalue 0 with an eigenspace of dimension $k-1$ and an eigenvalue 1 with an eigenspace of dimension 1. From this we see easily that the vector $\mathbf{q}$ is the only probability vector $\mathbf{x}$ that satisfies the equation $P\mathbf{x} = \mathbf{x}$. This vector $\mathbf{q}$ is called the **steady state vector** of the Markov process.

8.5.3.3 Finding the Steady State Vector

We list several methods for finding the steady state vector $\mathbf{q}$ of a given transition matrix P using *Scientific Notebook*.

1. Work out the matrix P^n numerically for some large values of n.
2. Point at the matrix $P - I$ and click on Matrices and Nullspace Basis. Then multiply the basis vector obtained by a suitable number to turn it into a probability vector.
3. Point at the matrix P and click on Matrices and Eigenvectors and select a probability vector corresponding to the eigenvalue 1.

8.5.3.4 Some Exercises

1. As an exercise, find the steady state vector of the transition matrix

$$\begin{bmatrix} .21357 & .12557 & .094629 & .10836 & .23951 & .12255 & .20681 \\ .1407 & .11187 & .16113 & .17647 & .14568 & .11029 & .054974 \\ .23367 & .21005 & .10997 & .19195 & .19012 & .16176 & .14136 \\ .027638 & .22603 & .15601 & .1548 & .02963 & .044118 & .081152 \\ .065327 & .14155 & .033248 & .14551 & .22469 & .1152 & .15969 \\ .10302 & .13242 & .23018 & .16409 & .046914 & .23039 & .21728 \\ .21608 & .052511 & .21483 & .058824 & .12346 & .21569 & .13874 \end{bmatrix}$$

Do the exercise by each of the suggested methods and verify that you obtain the same answer each time.

2. (a) Even though Theorem 8.5.3 is stated for transition matrices whose entries are all positive, verify numerically that the transition matrix

$$A = \begin{bmatrix} .2 & 0 & 0 & .1 & .2 & 0 & .1 \\ 0 & .1 & .4 & 0 & .3 & 0 & .2 \\ 0 & .3 & .3 & 0 & .4 & 0 & 0 \\ .3 & 0 & .2 & 0 & .1 & .4 & .3 \\ .1 & .2 & .1 & .2 & 0 & .3 & 0 \\ .4 & 0 & 0 & .3 & 0 & .2 & 0 \\ 0 & .4 & 0 & .4 & 0 & .1 & .4 \end{bmatrix}$$

does indeed have a steady state vector.

(b) Find an exact expression (not in decimal form) for the limit matrix and the steady state vector of the matrix A.

(c) Explore some matrices like the matrix A until you find one that does not have a steady state vector. Can you decide what sort of phenomenon has to be avoided in order for the steady state vector to exist?

3. Given that P is a regular transition matrix, prove that P has an eigenvalue 1 of multiplicity one and that every other eigenvalue (real or complex) of P must have absolute value less than 1. For a solution to this exercise, See item 1 of Appendix A.

Chapter 9
Sequences and Series

In this chapter we shall demonstrate several applications of the features of *Scientific Notebook* to the study of sequences and series of numbers and functions.

9.1 Some Recursively Defined Sequences

9.1.1 A Convergent Sequence

Suppose that a sequence (a_n) is defined recursively by the equations $a_1 = 3$ and

$$a_{n+1} = \sqrt{5 + a_n}.$$

It is not hard to prove by mathematical induction that this sequence is increasing and bounded above and that, consequently, it converges. Furthermore, if

$$a = \lim_{n \to \infty} a_n$$

then, letting $n \to \infty$ in the equation $a_{n+1} = \sqrt{5 + a_n}$ we obtain

$$a = \sqrt{5 + a}$$

Since this equation can be expressed in quadratic form we can solve it easily obtaining $a = \frac{1}{2} + \frac{1}{2}\sqrt{21}$.

We shall now look at some other ways in which the convergence of this sequence can be explored using *Scientific Notebook*.

9.1.1.1 Letting *Scientific Notebook* Solve the Equation

Point to the equation $a = \sqrt{5 + a}$ and click on Solve and Exact to obtain its solution. When solving more complicated problems it may be necessary to use Solve and Numeric.

9.1.1.2 The Fixed Point Method

We know that if $f(x) = \sqrt{x + 5}$ for all $x > -5$ then the number a must be a fixed point of this function f. To find this fixed point graphically we draw the graph of f and the line $y = x$ in the same system. Where these two graphs intersect is the required fixed point.

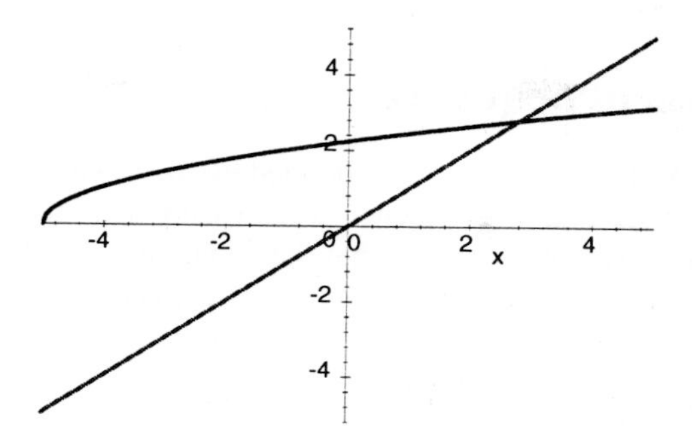

9.1.1.3 Looking at the Sequence

We point at the equation $f(x) = \sqrt{x+5}$ and inform *Scientific Notebook* of this definition by clicking on Define and New Definition. We now calculate a sequence of iterates of the function f starting at the number 3. Click on Calculus and Iterate ann we see the iterate dialog box.

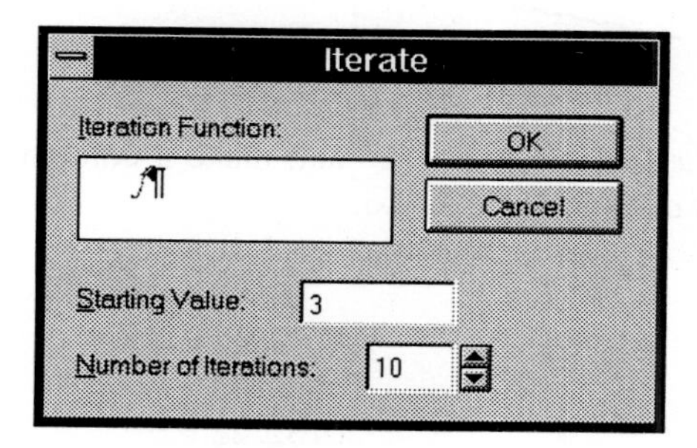

Fill it in as shown, typing f for the iteration function, 3 for the starting value and take the number of iterations as 10, and you will obtain

$$\text{Iterates:}\quad \begin{array}{c} 3 \\ 2.828427 \\ 2.797933 \\ 2.792478 \\ 2.791501 \\ 2.791326 \\ 2.791295 \\ 2.791289 \\ 2.791288 \\ 2.791288 \\ 2.791288 \end{array}$$

This table suggests very strongly that the limit of the given sequence is about 2.791288.

9.1.2 A Divergent Sequence

In this example we look at the sequence (a_n) defined by $a_1 = 3$ and $a_{n+1} = \sqrt{5 + a_n} + a_n^2$ for each n. It is easy to see that this sequence is increasing and that it is unbounded above. Consistent with this observation is the fact that the equation $a = \sqrt{5 + a} + a^2$ has no real solution. If we look for a fixed point of the function f defined by $f(x) = \sqrt{5 + x} + x^2$ by plotting the graph of f in the same system as the line $y = x$, then we obtain the next figure that shows that the two graphs do not intersect.

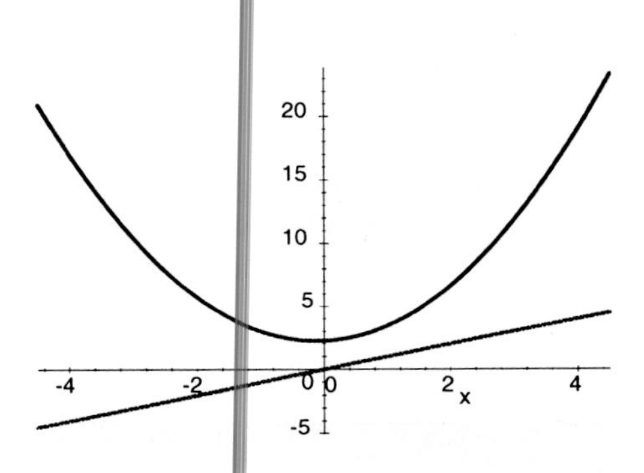

Finally, if we define $f(x) = \sqrt{5 + x} + x^2$ and make a column of iterates then we see compelling evidence that $a_n \to \infty$ as $n \to \infty$.

$$\text{Iterates:}\ \begin{array}{c} 3 \\ 11.82843 \\ 144.0139 \\ 20752.22 \\ 4.306548 \times 10^{8} \\ 1.854636 \times 10^{17} \\ 3.439673 \times 10^{34} \\ 1.183135 \times 10^{69} \\ 1.399809 \times 10^{138} \\ 1.959464 \times 10^{276} \\ 3.839501 \times 10^{552} \end{array}$$

9.1.3 Some Exercises

For each of the following recursively defined sequences, show that the sequence is monotone, try to decide whether or not it is bounded, try to find the limit yourself and then use the three methods shown above to let *Scientific Notebook* find the limit for you.

1. We define $a_1 = 2$ and for each $n \geq 1$ we have

$$a_{n+1} = \frac{1}{2}\left(a_n + \frac{2}{a_n}\right).$$

2. We define $a_1 = -1$ and for each $n \geq 1$ we have

$$a_{n+1} = \frac{1}{2}\left(a_n + \frac{2}{a_n}\right).$$

3. We define $a_1 = 1$ and for each $n \geq 1$ we have

$$a_{n+1} = \sqrt[3]{a_n^2 + a_n + 3}.$$

4. We define $a_1 = 1$ and for each $n \geq 1$ we have

$$a_{n+1} = \sqrt[3]{a_n^2 + \sqrt{a_n + 3}}.$$

5. We define $a_1 = 0$ and for each $n \geq 1$ we have

$$a_{n+1} = \sqrt[3]{\frac{6a_n + 1}{8}}.$$

Can you see why the limit of this sequence is $\cos\frac{\pi}{9}$? See item 2 of Appendix A.

9.2 Testing Series for Convergence

9.2.1 A Convergent *p*-Series

Using *Scientific Notebook* it is easy to verify that the series $\sum \frac{1}{n^2}$ converges. By pointing at the expression

$$\sum_{n=1}^{\infty} \frac{1}{n^2}$$

and clicking on Evaluate Numerically we obtain

$$\sum_{n=1}^{\infty} \frac{1}{n^2} = 1.644934.$$

We can, in fact, do better. By clicking on Evaluate we can see that

$$\sum_{n=1}^{\infty} \frac{1}{n^2} = \frac{\pi^2}{6}.$$

9.2.2 A Divergent Series

In this subsection we shall discuss the divergence of the series

$$\sum \frac{1}{(\log \log n)^{(\log \log n)}}.$$

The fact that this series does actually diverge can be seen from the comparison test. We begin by observing that since

$$\lim_{n \to \infty} \frac{(\log \log n)^2}{\log n} = 0$$

we have $(\log \log n)^2 < \log n$ for all n sufficiently large. Therefore, for n sufficiently large we have

$$(\log \log n)^{(\log \log n)} = \exp\left((\log \log n)^2\right) < \exp(\log n) = n$$

and so the given series can be compared with the divergent series $\sum \frac{1}{n}$.

We shall now explore the divergence of this series with *Scientific Notebook* and see how it could have been anticipated. For each integer $n \geq 4$ we define

$$f(n) = \sum_{j=8}^{n} \frac{1}{(\log \log j)^{(\log \log j)}}$$

and we inform *Scientific Notebook* of this definition by clicking on Define and New Definition. We can now sum the series to any specified number of terms be evaluating the function f numerically at that number. For example, by pointing at the expression $f(1000)$ and clicking on Evaluate Numerically, we obtain

$$f(1000) = 373.5185.$$

An even more revealing view of the function f can be obtained if we evaluate it at on a whole column of numbers. Click on the matrix button that is in your Math toolbar and select a matrix with one column and 10 rows. Fill in the matrix as follows

$$\begin{bmatrix} 100 \\ 200 \\ 400 \\ 600 \\ 800 \\ 1000 \\ 1200 \\ 1400 \\ 1600 \end{bmatrix}.$$

When you evaluate the function f on this matrix you will obtain

$$f\begin{bmatrix}100\\200\\400\\600\\800\\1000\\1200\\1400\\1600\end{bmatrix}=\begin{bmatrix}64.93046\\111.6022\\188.2803\\254.9336\\316.0698\\373.5185\\428.2516\\480.8633\\531.7506\end{bmatrix}.$$

In this way we can explore the divergence of the given series and obtain a sense of how rapidly the partial sum tends to infinity.

9.2.3 Another Divergent Series

In this subsection we shall discuss the series

$$\sum\frac{(4^n)(n!)^2}{(2n)!}.$$

As in Example 9.2.2 we can anticipate the fact that the series diverges by looking at the partial sums: For each integer $n\geq 1$ we define

$$f(n)=\sum_{j=1}^{n}\frac{(4^j)(j!)^2}{(2j)!}$$

and we inform *Scientific Notebook* of this definition by pointing at the latter equation and clicking on Define and New Definition. We can then obtain

$$f\begin{bmatrix}100\\200\\400\\600\\800\\1000\\1200\\1400\\1600\end{bmatrix}=\begin{bmatrix}1194.278\\3360.315\\9479.015\\17398.33\\26774.31\\37407.98\\49165.07\\61946.95\\75677.21\end{bmatrix}.$$

This view of the partial sums makes a very strong case for the divergence of the given series. In fact, it is clear that this series diverges much more rapidly than the one we studied in Example 9.2.2.

We shall now see how *Scientific Notebook* can be used to give a rigorous proof that the given series diverges. We begin by defining

$$a_n = \frac{(4^n)(n!)^2}{(2n)!}$$

and we inform *Scientific Notebook* of this definition by clicking on Define and New Definition. The Interpret Subscript dialog box

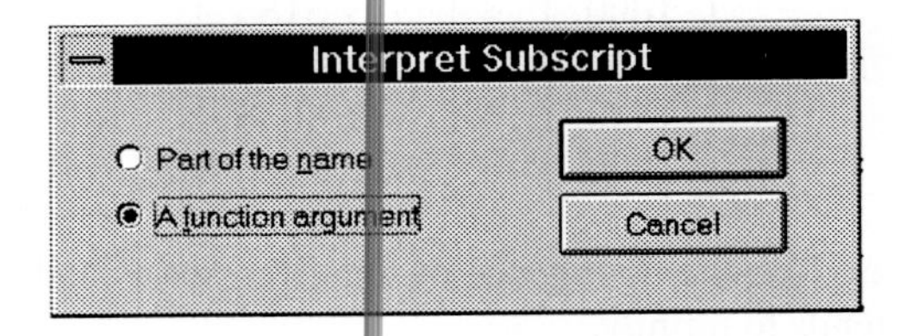

appears and we inform *Scientific Notebook* that the subscript n in the expression a_n is a function argument. We now point at the ratio $\frac{a_{n+1}}{a_n}$ and click on Simplify to obtain

$$\frac{a_{n+1}}{a_n} = 2\frac{n+1}{2n+1}$$

from which we deduce that

$$\frac{a_{n+1}}{a_n} > 1$$

for each n. Since the sequence (a_n) is increasing, it cannot approach 0 and so $\sum a_n$ is divergent.

9.2.4 Testing a Series with the Ratio Test

The **d'Alembert ratio test** says that if $a_n > 0$ for every n and if

$$\lim_{n\to\infty} \frac{a_{n+1}}{a_n} = r$$

then the series $\sum a_n$ will converge if $r < 1$ and will diverge if $r > 1$. In this subsection we shall consider the series $\sum a_n$ where for each n we have

$$a_n = (4e)^{\frac{n(n+1)}{2}} \frac{(1^1)(1!)(2^2)(2!)\cdots(n^n)(n!)}{\left(3^{2+\frac{1}{2}}\right)\left(5^{4+\frac{1}{2}}\right)\cdots\left((2n+1)^{2n+\frac{1}{2}}\right)}$$

which we can also write in the form

$$a_n = (4e)^{\frac{n(n+1)}{2}} \prod_{j=1}^{n} \frac{(j^j)(j!)}{(2j+1)^{2j+\frac{1}{2}}}.$$

If we point at the latter equation and click on Define and New Definition and then point at the expression $\lim_{n\to\infty} \frac{a_{n+1}}{a_n}$ and click on Evaluate Numerically we obtain

$$\lim_{n\to\infty} \frac{a_{n+1}}{a_n} = .6520493.$$

The given series therefore diverges. If we look a little more closely at the expression $\frac{a_{n+1}}{a_n}$ we see that it is given by

$$\frac{a_{n+1}}{a_n} = \frac{(n^n)(n!)(4e)^n}{\left((2n+1)^{\left(2n+\frac{1}{2}\right)}\right)}$$

and so we can obtain a precise value for $\lim_{n\to\infty} \frac{a_{n+1}}{a_n}$ by using Evaluate to show that

$$\lim_{n\to\infty} \frac{(n^n)(n!)(4e)^n}{\left((2n+1)^{\left(2n+\frac{1}{2}\right)}\right)} = \frac{\sqrt{\pi}}{e}.$$

If we look at the limit this way then we can deduce the divergence of the given series from the fact that $\sqrt{\pi} < e$.

9.2.5 Testing a Series with Raabe's Test

The standard ratio test, more properly called d'Alembert's test, gives no information about a given series $\sum a_n$ if

$$\lim_{n\to\infty} \frac{a_{n+1}}{a_n} = 1.$$

However, a sharper version of the ratio test, known as **Raabe's test**, may still provide us with a conclusion. Raabe's test says that if $a_n > 0$ for each n and if

$$\lim_{n\to\infty} n\left(1 - \frac{a_{n+1}}{a_n}\right) = p$$

then the series $\sum a_n$ will converge if $p > 1$ and will diverge if $p < 1$.

In this subsection we shall look at a variation of the series that we studied in Subsection 9.2.4. We shall consider the series $\sum a_n$ where

$$a_n = \left(\frac{\pi^{\frac{n}{2}}}{2^{n(n+1)}e^{\frac{n(n+3)}{2}}}\right) \frac{\left(3^{2+\frac{1}{2}}\right)\left(5^{4+\frac{1}{2}}\right)\cdots\left((2n+1)^{2n+\frac{1}{2}}\right)}{(1^1)(1!)(2^2)(2!)\cdots(n^n)(n!)}$$

which we can also write in the form

$$a_n = \left(\frac{\pi^{\frac{n}{2}}}{2^{n(n+1)}e^{\frac{n(n+3)}{2}}}\right) \prod_{j=1}^{n} \frac{(2j+1)^{2j+\frac{1}{2}}}{(j^j)(j!)}.$$

Once again, we give this definition to *Scientific Notebook*. This time, however, if we point to the expression $\lim_{n\to\infty} \frac{a_{n+1}}{a_n}$ and click on Evaluate Numerically we see that

$$\lim_{n\to\infty} \frac{a_{n+1}}{a_n} = 1.0$$

and so the d'Alembert version of the ratio test doesn't help us. We can also obtain this limit precisely by observing that for each n we have

$$\frac{a_{n+1}}{a_n} = \frac{\left((2n+3)^{\left(2n+\frac{5}{2}\right)}\right)\sqrt{\pi}}{\left((n+1)^{n+1}\right)((n+1)!)\,2^{2n+2}e^{n+2}}$$

and by clicking on Evaluate we see that

$$\lim_{n\to\infty} \frac{a_{n+1}}{a_n} = \lim_{n\to\infty} \frac{\left((2n+3)^{\left(2n+\frac{5}{2}\right)}\right)\sqrt{\pi}}{\left((n+1)^{n+1}\right)((n+1)!)\,2^{2n+2}e^{n+2}} = 1.$$

On the other hand,

$$\lim_{n\to\infty} n\left(1 - \frac{a_{n+1}}{a_n}\right) = \lim_{n\to\infty} n\left(1 - \frac{\left((2n+3)^{\left(2n+\frac{5}{2}\right)}\right)\sqrt{\pi}}{\left((n+1)^{n+1}\right)((n+1)!)\,2^{2n+2}e^{n+2}}\right) = \frac{1}{12}.$$

Since this limit is less than 1 we conclude from Raabe's test that the given series diverges.

9.2.6 Further Remarks about the Preceding Series

As in Subsection 9.2.5 we define

$$a_n = \left(\frac{\pi^{\frac{n}{2}}}{2^{n(n+1)}e^{\frac{n(n+3)}{2}}}\right)\prod_{j=1}^{n} \frac{(2j+1)^{2j+\frac{1}{2}}}{(j^j)(j!)}$$

for each n. We have seen that the series $\sum a_n$ diverges but this leaves open the question as to whether or not $a_n \to 0$ as $n \to \infty$. We can answer this question using the following theorem.

Suppose that (a_n) *is a decreasing sequence of positive numbers and that, for each* n *we have*

$$b_n = 1 - \frac{a_{n+1}}{a_n}.$$

Then the sequence (a_n) *approaches zero as* $n \to \infty$ *if and only if the series* $\sum b_n$ *diverges.*[18]

We recall that for the series presently being discussed,

$$\lim_{n\to\infty} nb_n = \lim_{n\to\infty} n\left(1 - \frac{a_{n+1}}{a_n}\right) = \frac{1}{12}$$

---

[18] This theorem may be found as Theorem 8.7.10 in the book *An Introduction to Mathematical Analysis* by Jonathan Lewin and Myrtle Lewin, (McGraw-Hill, 1988.)

and so it follows by comparison of $\sum b_n$ with the divergent series $\sum \frac{1}{n}$ that $\sum b_n$ diverges. We conclude that, even though $\sum a_n$ diverges, we still have

$$\lim_{n \to \infty} a_n = 0.$$

Interestingly enough, neither Maple nor Mathematica seem to be willing to evaluate this limit directly.

9.3 Rate of Convergence

In this section we look at some ways in which *Scientific Notebook* can be used to compare the rates at which two given series converge.

9.3.1 Example of Two Convergent Series

In this example, we shall consider the series $\sum \frac{1}{n2^n}$ and $\sum \frac{(-1)^{n-1}}{n}$. It is not hard to show that

$$\sum_{n=1}^{\infty} \frac{1}{n2^n} = \ln 2 = \sum_{n=1}^{\infty} \frac{(-1)^{n-1}}{n}.$$

In order to compare the rates of convergence of these two series, we define

$$f(n) = \sum_{j=1}^{n} \frac{1}{j2^j}$$

and

$$g(n) = \sum_{j=1}^{n} \frac{(-1)^{j-1}}{j}$$

for each positive integer n and we inform *Scientific Notebook* of these definitions by pointing at the equations and clicking on Define and New Definition. Then we point at the matrix

$$\begin{array}{cc} 1 & f(1) \\ 2 & f(2) \\ 3 & f(3) \\ 4 & f(4) \\ 5 & f(5) \\ 6 & f(6) \\ 7 & f(7) \\ 8 & f(8) \\ 9 & f(9) \\ 10 & f(10) \\ 11 & f(11) \\ 12 & f(12) \\ 13 & f(13) \\ 14 & f(14) \\ 15 & f(15) \end{array}$$

and click on Plot2D and Rectangular and drag each of the expressions $\ln 2$ and

$$\begin{array}{cc} 1 & g(1) \\ 2 & g(2) \\ 3 & g(3) \\ 4 & g(4) \\ 5 & g(5) \\ 6 & g(6) \\ 7 & g(7) \\ 8 & g(8) \\ 9 & g(9) \\ 10 & g(10) \\ 11 & g(11) \\ 12 & g(12) \\ 13 & g(13) \\ 14 & g(14) \\ 15 & g(15) \end{array}$$

into our plot. In this way we obtain the next figure. In this figure, the graph of f is shown in blue and the graph of g is shown in red. We see at once that the series $\sum \frac{1}{n2^n}$ converges much more rapidly than the series $\sum \frac{(-1)^{n-1}}{n}$.

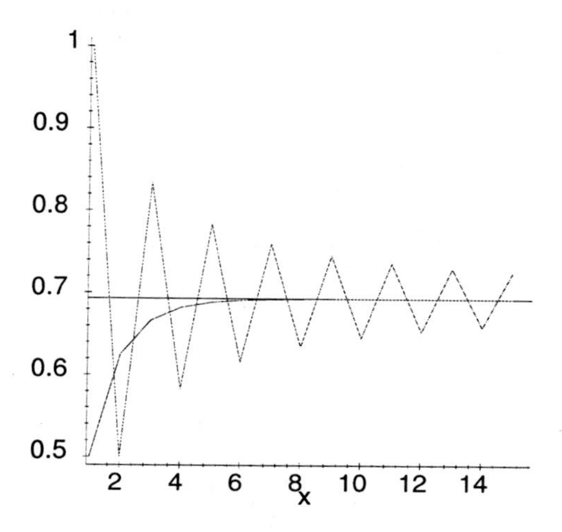

9.3.2 Exercise

Use a graphical method to compare the rates of convergence of the series

$$\sum \frac{(-1)^{n-1}(2n)!}{4^n (n!)^2} \quad \text{and} \quad \sum \frac{\sqrt{2}(-1)^{n-1}}{(\sqrt{2}-1)\, n \log 2}.$$

9.4 Power Series

9.4.1 Brief Introduction

A series of the form

$$\sum a_n x^n$$

is called a **power series** in the symbol x with coefficients (a_n). One of the important facts about power series is the following theorem:

Suppose that $r > 0$ and that

$$f(x) = \sum_{n=0}^{\infty} a_n x^n$$

whenever $-r < x < r$. Then whenever $-r < x < r$ we have

$$f'(x) = \sum_{n=1}^{\infty} n a_n x^{n-1}.$$

We usually sum up the statement of this theorem by saying that a power series can be differentiated **term by term**. Repeating this process k times for any natural number k we see that the equation

$$f^{(k)}(x) = \sum_{n=k}^{\infty} n(n-1)(n-2)\cdots(n-k+1)\, a_n x^{n-k}$$

whenever $-r < x < r$ and, putting $x = 0$ in the latter equation we obtain

$$f^{(k)}(0) = (k!)\, a_k.$$

This information can be summed up in the following theorem:

Suppose that $r > 0$ and that

$$f(x) = \sum_{n=0}^{\infty} a_n x^n$$

whenever $-r < x < r$. Then for every n we have $a_n = \frac{f^{(n)}(0)}{n!}$ and so, for every $x \in (-r, r)$ we have

$$f(x) = \sum_{n=0}^{\infty} \frac{f^{(n)}(0)}{n!} x^n.$$

Given any function that has derivatives of all orders at the number 0, the series

$$\sum \frac{f^{(n)}(0)}{n!} x^n$$

is called the **Maclaurin series** of the function f. We see that one of the following two conditions must hold:

1. For every number x in some open interval that contains 0 we have

$$f(x) = \sum_{n=0}^{\infty} \frac{f^{(n)}(0)}{n!} x^n.$$

2. It is impossible to find a power series $\sum a_n x^n$ for which the equation

$$f(x) = \sum_{n=0}^{\infty} a_n x^n$$

holds for all numbers x in an open interval containing 0.

A key question to ask about any given function f is whether it is possible to find an interval in which the Maclaurin series of f converges to f, and for what intervals this will happen.

9.4.2 The Maclaurin Polynomials of a Function

Once again we suppose that f is a function that has derivatives of all orders on an open interval containing 0. For each nonnegative integer n, the nth partial sum of the Maclaurin series of f is called the nth **Maclaurin polynomial** of f. If we call this polynomial f_n then for each number x we have

$$f_n(x) = \sum_{j=0}^{n-1} \frac{f^{(j)}(0)}{j!} x^j.$$

The condition for f to be the sum of its Maclaurin series at a given number x is that $f_n(x) \to f(x)$ as $n \to \infty$.

9.4.3 Calculating Maclaurin Polynomials with *Scientific Notebook*

We shall illustrate the method of calculating a Maclaurin polynomial by showing how to calculate the polynomial f_{23} where the function f is defined to be

$$f(x) = x \sin\left(x^2 + 3x + 1\right)$$

for each number x. The first thing we have to do is inform *Scientific Notebook* of this definition by pointing at the latter equation and clicking on Define and New Definition. We now proceed as follows:

1. Point at the expression $f(x)$, open the Maple menu and click on Power Series to bring up the series expansion dialog box

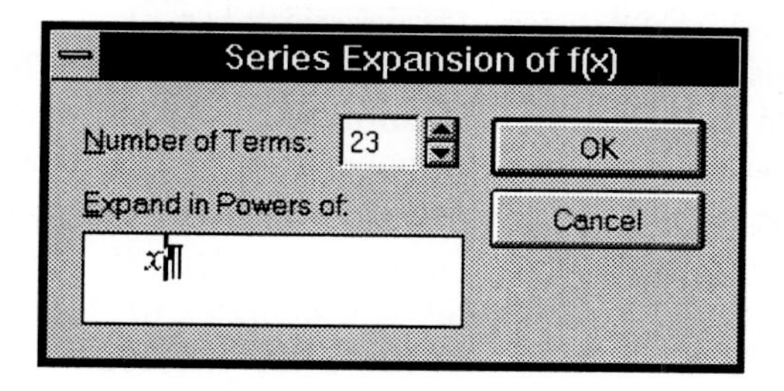

which we fill as shown with the number of terms declared as 23 and the name of the variable declared as x. Click on OK.

2. The expression before you consists of a single long line and, because the number of terms is so large, this line disappears into the right margin. Place your cursor to the left of this long expression, hold down your shift key and press the end key to highlight the entire line. Now, while holding down the control key, click on Evaluate. The result will now appear with enough line breaks to prevent it from disappearing into the right margin and you will see:[19]

---

[19] The following expression will be fully visible only in the on-screen version of this text.

$$
\begin{aligned}
&f(x) = (\sin 1)\,x + 3(\cos 1)\,x^2 + \left(-\tfrac{9}{2}\sin 1 + \cos 1\right)x^3 + \left(-\tfrac{9}{2}\cos 1 - 3\sin 1\right)x^4 + \\
&\left(-\tfrac{9}{2}\cos 1 + \tfrac{23}{8}\sin 1\right)x^5 + \left(\tfrac{9}{2}\sin 1 + \tfrac{21}{40}\cos 1\right)x^6 + \left(\tfrac{99}{80}\sin 1 + \tfrac{77}{24}\cos 1\right)x^7 + \\
&\left(\tfrac{1017}{560}\cos 1 - \tfrac{61}{40}\sin 1\right)x^8 + \left(-\tfrac{21}{80}\cos 1 - \tfrac{19933}{13440}\sin 1\right)x^9 + \left(-\tfrac{177}{560}\sin 1 - \tfrac{3733}{4480}\cos 1\right)x^{10} + \\
&\left(\tfrac{13551}{44800}\sin 1 - \tfrac{5261}{13440}\cos 1\right)x^{11} + \left(\tfrac{12333}{492800}\cos 1 + \tfrac{1157}{4480}\sin 1\right)x^{12} + \left(\tfrac{5151}{44800}\cos 1 + \tfrac{149489}{2534400}\sin 1\right)x^{13} + \\
&\left(\tfrac{4100903}{76876800}\cos 1 - \tfrac{2139}{70400}\sin 1\right)x^{14} + \left(-\tfrac{9993647}{358758400}\sin 1 + \tfrac{33983}{17740800}\cos 1\right)x^{15} + \\
&\left(-\tfrac{576239}{76876800}\sin 1 - \tfrac{17260543}{1793792000}\cos 1\right)x^{16} + \left(-\tfrac{350683}{71751680}\cos 1 + \tfrac{404185601}{258306048000}\sin 1\right)x^{17} + \\
&\left(-\tfrac{80599069}{1330667552000}\cos 1 + \tfrac{325253}{163072000}\sin 1\right)x^{18} + \left(\tfrac{628430951}{9758228480000}\sin 1 + \tfrac{1176986729}{23247544320000}\cos 1\right)x^{19} + \\
&\left(-\tfrac{213166249}{131736084480000}\sin 1 + \tfrac{825417251}{26486620160000}\cos 1\right)x^{20} + \left(\tfrac{77753033}{12546293760000}\cos 1 - \tfrac{2924701597819}{300358272614400000}\sin 1 \right. \\
&\left(-\tfrac{1082904765433}{700835969433600000}\cos 1 - \tfrac{6429811133}{16686570700800 0}\sin 1\right)x^{22} + Ox^{23}
\end{aligned}
$$

3. The last term that appears as Ox^{23} is simply $f(x) - f_{23}(x)$. This notation is a hint that for "nice" functions f, the expression

$$\frac{f(x) - f_n(x)}{x^n}$$

should be bounded as $n \to \infty$. Delete the term Ox^{23} and change the left side from $f(x)$ to $f_{23}(x)$, and you have calculated $f_{23}(x)$ and you should see[20]

$$
\begin{aligned}
&f_{23}(x) = (\sin 1)\,x + 3(\cos 1)\,x^2 + \left(-\tfrac{9}{2}\sin 1 + \cos 1\right)x^3 + \left(-\tfrac{9}{2}\cos 1 - 3\sin 1\right)x^4 + \\
&\left(-\tfrac{9}{2}\cos 1 + \tfrac{23}{8}\sin 1\right)x^5 + \left(\tfrac{9}{2}\sin 1 + \tfrac{21}{40}\cos 1\right)x^6 + \left(\tfrac{99}{80}\sin 1 + \tfrac{77}{24}\cos 1\right)x^7 + \\
&\left(\tfrac{1017}{560}\cos 1 - \tfrac{61}{40}\sin 1\right)x^8 + \left(-\tfrac{21}{80}\cos 1 - \tfrac{19933}{13440}\sin 1\right)x^9 + \left(-\tfrac{177}{560}\sin 1 - \tfrac{3733}{4480}\cos 1\right)x^{10} + \\
&\left(\tfrac{13551}{44800}\sin 1 - \tfrac{5261}{13440}\cos 1\right)x^{11} + \left(\tfrac{12333}{492800}\cos 1 + \tfrac{1157}{4480}\sin 1\right)x^{12} + \left(\tfrac{5151}{44800}\cos 1 + \tfrac{149489}{2534400}\sin 1\right)x^{13} + \\
&\left(\tfrac{4100903}{76876800}\cos 1 - \tfrac{2139}{70400}\sin 1\right)x^{14} + \left(-\tfrac{9993647}{358758400}\sin 1 + \tfrac{33983}{17740800}\cos 1\right)x^{15} + \\
&\left(-\tfrac{576239}{76876800}\sin 1 - \tfrac{17260543}{1793792000}\cos 1\right)x^{16} + \left(-\tfrac{350683}{71751680}\cos 1 + \tfrac{404185601}{258306048000}\sin 1\right)x^{17} + \\
&\left(-\tfrac{80599069}{1330667552000}\cos 1 + \tfrac{325253}{163072000}\sin 1\right)x^{18} + \left(\tfrac{628430951}{9758228480000}\sin 1 + \tfrac{1176986729}{23247544320000}\cos 1\right)x^{19} + \\
&\left(-\tfrac{213166249}{131736084480000}\sin 1 + \tfrac{825417251}{26486620160000}\cos 1\right)x^{20} + \left(\tfrac{77753033}{12546293760000}\cos 1 - \tfrac{2924701597819}{300358272614400000}\sin 1 \right. \\
&\left(-\tfrac{1082904765433}{700835969433600000}\cos 1 - \tfrac{6429811133}{1668657070 08000}\sin 1\right)x^{22}.
\end{aligned}
$$

4. You may also wish to evaluate the function f_{23} numerically. To do this, highlight the expression on the right side, hold down the control button and click on Evaluate Numerically, You will now obtain the polynomial $f_{23}(x)$ in the form

$$
\begin{aligned}
&f_{23}(x) = .841471x + 1.620907x^2 - 3.246317x^3 - 4.955773x^4 - .0121313x^5 + \\
&4.070278x^6 + 2.77479x^7 - .3020157x^8 - 1.389823x^9 - .7161767x^{10} + 4.302839 \times \\
&10^{-2}x^{11} + .2308392x^{12} + .111756x^{13} + 3.254942 \times 10^{-3}x^{14} - 2.240522 \times 10^{-2}x^{15} - \\
&1.150634 \times 10^{-2}x^{16} - 1.324007 \times 10^{-3}x^{17} + 1.351081 \times 10^{-3}x^{18} + 8.154548 \times \\
&10^{-4}x^{19} + 1.547613 \times 10^{-4}x^{20} - 4.845309 \times 10^{-5}x^{21} - 4.077282 \times 10^{-5}x^{22}
\end{aligned}
$$

9.4.4 Graphing a Maclaurin Polynomial

In this subsection we shall return to the function f defined in Subsection 9.4.3 and show graphically how well the polynomial f_{23} approximates f in a given interval. We begin by pointing at the expression $f(x)$ and clicking on Plot 2D and Rectangular. In a moment we are going to drag the graph $y = f_{23}(x)$ into the sketch that we have made, but first we have

---

[20] The following expression will be fully visible only in the on-screen version of this text.

$10^{-2}\cos 5.0x + 1.783987 \times 10^{-2}\sin 5.0x - 2.147171 \times 10^{-2}\cos 6.0x + 1.076326 \times 10^{-2}\sin 6.0x - 1.599178 \times 10^{-2}\cos 7.0x + 1.773685 \times 10^{-3}\sin 7.0x - .0104423\cos 8.0x + 2.210998 \times 10^{-3}\sin 8.0x - 1.110025 \times 10^{-2}\cos 9.0x + 1.85077 \times 10^{-3}\sin 9.0x - 9.20957 \times 10^{-3}\cos 10.0x - 1.337288 \times 10^{-3}\sin 10.0x$

9.5.4 Faster Calculation of Fourier Polynomials

Generally the rule is that, in any long and complex computation, the fewer times the computer has to call upon a function using the definition you have given it with Define and New Definition, the faster the computation will be. Thus you can speed up the computation of Fourier polynomials dramatically by writing in the functions directly. In some vintages of *Scientific Notebook*, writing the functions directly is also more reliable. As before, we start off by pointing at the definition

$$f(x) = \left|\frac{x \log \frac{x}{2}}{1+x^2}\right|$$

and clicking on Define and New Definition. Now, to avoid calling upon definitions of the coefficients a_n and b_n we define $p(n,x)$ using their explicit values. In other words, we express $p(n,x)$ in the form

$$p(n,x) = \frac{1}{2\pi}\int_0^{2\pi} f(t)\,dt + \frac{1}{\pi}\sum_{j=1}^{n}\left(\left(\int_0^{2\pi} f(t)\cos jt dt\right)\cos jx + \left(\int_0^{2\pi} f(t)\sin jt dt\right)\sin jx\right).$$

You now have a choice. Either you can send this definition of $p(n,x)$ to *Scientific Notebook* and then evaluate an expression like $p(10,x)$ by pointing at it and clicking on Evaluate Numerically, or you can copy the preceding formula for $p(n,x)$ and change the symbol n to 10 on each side. You will have

$$p(10,x) = \frac{1}{2\pi}\int_0^{2\pi} f(t)\,dt + \frac{1}{\pi}\sum_{j=1}^{10}\left(\left(\int_0^{2\pi} f(t)\cos jt dt\right)\cos jx + \left(\int_0^{2\pi} f(t)\sin jt dt\right)\sin jx\right)$$

and you can calculate $p(10,x)$ by highlighting the right side, holding down the control key and clicking on Evaluate Numerically. Once again you will obtain

$p(10,x) = .1924698 + .1139326\cos x + 1.475627 \times 10^{-2}\sin x + 3.914197 \times 10^{-2}\cos 2.0x + .1056249\sin 2.0x - 3.226529 \times 10^{-2}\cos 3.0x + 6.431681 \times 10^{-2}\sin 3.0x - 2.723772 \times 10^{-2}\cos 4.0x + 2.289696 \times 10^{-2}\sin 4.0x - 1.852134 \times 10^{-2}\cos 5.0x + 1.783987 \times 10^{-2}\sin 5.0x - 2.147171 \times 10^{-2}\cos 6.0x + 1.076326 \times 10^{-2}\sin 6.0x - 1.599178 \times 10^{-2}\cos 7.0x + 1.773685 \times 10^{-3}\sin 7.0x - .0104423\cos 8.0x + 2.210998 \times 10^{-3}\sin 8.0x - 1.110025 \times 10^{-2}\cos 9.0x + 1.85077 \times 10^{-3}\sin 9.0x - 9.20957 \times$

$10^{-3}\cos 10.0x - 1.337288 \times 10^{-3}\sin 10.0x$

but this time the calculation will be much faster than it was in Subsection 9.5.3.

An even faster way of evaluating $p(10, x)$ is to avoid the Define operation completely. In the formula

$$\frac{1}{2\pi}\int_0^{2\pi} f(t)\,dt + \frac{1}{\pi}\sum_{j=1}^{n}\left(\left(\int_0^{2\pi} f(t)\cos jt\,dt\right)\cos jx + \left(\int_0^{2\pi} f(t)\sin jt\,dt\right)\sin jx\right)$$

for $p(n, x)$, replace the expression $f(t)$ by its actual value. In other words, write $p(n, x)$ in the form

$$\begin{aligned} p(n,x) &= \frac{1}{2\pi}\int_0^{2\pi}\left|\frac{t\log\frac{t}{2}}{1+t^2}\right| dt + \\ &\quad \frac{1}{\pi}\sum_{j=1}^{n}\left(\left(\int_0^{2\pi}\left|\frac{t\log\frac{t}{2}}{1+t^2}\right|\cos jt\,dt\right)\cos jx + \left(\int_0^{2\pi}\left|\frac{t\log\frac{t}{2}}{1+t^2}\right|\sin jt\,dt\right)\sin jx\right). \end{aligned}$$

To calculate $p(10, x)$, copy this explicit formula for $p(n, x)$, replace the symbol n by 10 and click on Evaluate Numerically.

9.5.5 Graphing a Fourier Polynomial

In this subsection we shall return to the function f defined in Subsection 9.5.3 and show graphically how well the polynomial $p(10, x)$ approximates $f(x)$ for $x \in [0, 2\pi)$. We begin by pointing at the expression $f(x)$ and clicking on Plot 2D and Rectangular. Open the plot properties dialog box, move to the page Plot Components and set the domain interval as $[0, 6.283]$. Now highlight the expression that we obtained in Subsection 9.5.3 for $p(10, x)$ and drag it into the sketch. Open the plot properties dialog box again, move to the page Plot Components, go to Item Number 2, set the color as red. You will now see the two graphs as shown in the next figure. In this figure, the graph $y = f(x)$ is shown in blue and the graph $y = p(10, x)$ is shown in red. Notice how closely the Fourier polynomial approximates the given function, except near the endpoints of the interval.

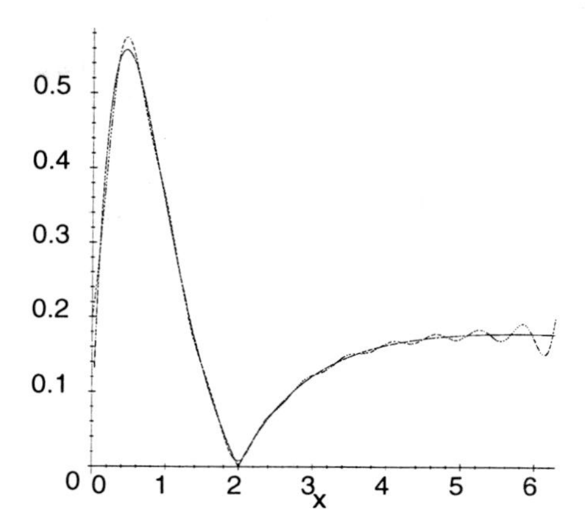

To see an animated view of the Fourier polynomials $p(n, x)$ as n ranges from 1 to 10, **click here**.

9.5.6 Fourier Polynomials at a Discontinuity

From a result known as **Dirichlet's theorem**, we know that if a given function f is either increasing or decreasing in a neighborhood of a given number x then the Fourier series of f converges at the number x to

$$\frac{1}{2}\left(\lim_{t\to x-} f(t) + \lim_{t\to x+} f(t)\right).$$

However, unless f is continuous at a given number x, the Fourier polynomials of f will be rather poor approximations to the function near the number x. This latter phenomenon is known as the **Gibbs phenomenon.** In this subsection, we shall illustrate these two theorems by showing the graphs of Fourier polynomials of a discontinuous function. We define

$$f(x) = x\left(\frac{x-\pi}{|x-\pi|}\right)$$

for $x \in [0, 2\pi)$. Note that

$$f(x) = \begin{cases} -x & \text{if} \quad 0 \le x < \pi \\ x & \text{if} \quad \pi < x < 2\pi \end{cases}$$

and that f fails to be continuous at the number π. The Fourier polynomial is, as usual, given by

$$p(n, x) = \frac{1}{2\pi}\int_0^{2\pi} f(t)\,dt +$$

$$\frac{1}{\pi}\sum_{j=1}^{n}\left(\left(\int_0^{2\pi} f(t)\cos jtdt\right)\cos jx + \left(\int_0^{2\pi} f(t)\sin jtdt\right)\sin jx\right).$$

To evaluate $p(n,x)$ for a given value of n we can use any of the following methods:

1. We can supply the definition of the function f to *Scientific Notebook* and than point at the latter formula for $p(n,x)$ and click on Evaluate Numerically.
2. We can write $p(n,x)$ more explicitly as

$$\begin{aligned} p(n,x) &= \frac{1}{2\pi}\int_0^{2\pi} t\left(\frac{t-\pi}{|t-\pi|}\right)dt + \\ &\frac{1}{\pi}\sum_{j=1}^{n}\left(\left(\int_0^{2\pi} t\left(\frac{t-\pi}{|t-\pi|}\right)\cos jtdt\right)\cos jx + \left(\int_0^{2\pi} t\left(\frac{t-\pi}{|t-\pi|}\right)\sin jtdt\right)\sin jx\right). \end{aligned}$$

3. We can split the interval of integration into the two parts $[0,\pi]$ and $[\pi,2\pi]$ and obtain $p(n,x)$ as

$$-\frac{1}{2\pi}\int_0^{\pi} tdt - \frac{1}{\pi}\sum_{j=1}^{n}\left(\left(\int_0^{\pi} t\cos jtdt\right)\cos jx + \left(\int_0^{\pi} t\sin jtdt\right)\sin jx\right) +$$

$$\frac{1}{2\pi}\int_\pi^{2\pi} tdt + \frac{1}{\pi}\sum_{j=1}^{n}\left(\left(\int_\pi^{2\pi} t\cos jtdt\right)\cos jx + \left(\int_\pi^{2\pi} t\sin jtdt\right)\sin jx\right)$$

This is the most efficient way to calculate $p(n,x)$.

So, for example, to obtain $p(30,x)$ we can highlight the expression

$$-\frac{1}{2\pi}\int_0^{\pi} tdt - \frac{1}{\pi}\sum_{j=1}^{30}\left(\left(\int_0^{\pi} t\cos jtdt\right)\cos jx + \left(\int_0^{\pi} t\sin jtdt\right)\sin jx\right) +$$

$$\frac{1}{2\pi}\int_\pi^{2\pi} tdt + \frac{1}{\pi}\sum_{j=1}^{30}\left(\left(\int_\pi^{2\pi} t\cos jtdt\right)\cos jx + \left(\int_\pi^{2\pi} t\sin jtdt\right)\sin jx\right),$$

hold down the control key and click on Evaluate Numerically to obtain

$1.27324\cos x - 4.0\sin x - 1.333333\sin 3.0x - .2105263\sin 19.0x+$
$4.135622\times 10^{-13}\cos 4.0x + 5.658842\times 10^{-3}\cos 15.0x + 4.13077\times 10^{-13}\cos 24.0x +$
$1.052264\times 10^{-2}\cos 11.0x + 5.092958\times 10^{-2}\cos 5.0x + 7.533962\times 10^{-3}\cos 13.0x -$
$.1481481\sin 27.0x + 4.127727\times 10^{-13}\cos 6.0x - .3076923\sin 13.0x+$
$2.40688\times 10^{-3}\cos 23.0x + 2.598448\times 10^{-2}\cos 7.0x + 4.162267\times 10^{-13}\cos 8.0x +$
$1.746556\times 10^{-3}\cos 27.0x - .2666667\sin 15.0x + 1.571901\times 10^{-2}\cos 9.0x + 1.570796-$
$.173913\sin 23.0x + 4.116542\times 10^{-13}\cos 10.0x - .3636364\sin 11.0x + 4.166544\times$
$10^{-13}\cos 14.0x + 4.142201\times 10^{-13}\cos 2.0x - .4444444\sin 9.0x+$
$4.110786\times 10^{-13}\cos 20.0x + 1.513959\times 10^{-3}\cos 29.0x + .1414711\cos 3.0x + 4.02756\times$
$10^{-13}\cos 28.0x + 4.130688\times 10^{-13}\cos 12.0x + 3.526979\times 10^{-3}\cos 19.0x - .8\sin 5.0x+$

$4.123533 \times 10^{-13} \cos 16.0x - .5714286 \sin 7.0x - .137931 \sin 29.0x+$
$2.037183 \times 10^{-3} \cos 25.0x + 2.887165 \times 10^{-3} \cos 21.0x + 4.405673 \times 10^{-3} \cos 17.0x -$
$.16 \sin 25.0x + 4.297183 \times 10^{-13} \cos 30.0x + 4.138029 \times 10^{-13} \cos 22.0x + 4.106198 \times$
$10^{-13} \cos 26.0x - .1904762 \sin 21.0x - .2352941 \sin 17.0x + 4.115747 \times 10^{-13} \cos 18.0x.$

To illustrate the polynomial $p(30, x)$, point at the expression $f(x)$ and click on Plot 2D and Rectangular. Open the plot properties dialog box, move to the page Plot Components and set the domain interval as $[0, 6.283]$. Now highlight the expression that we obtained for $p(30, x)$ and drag it into the sketch. Open the plot properties dialog box again, move to the page Plot Components, go to Item Number 2, set the color as red. You will now see the two graphs as shown in the next figure. In this figure, the graph $y = f(x)$ is shown in blue and the graph $y = p(30, x)$ is shown in red. Notice how closely the Fourier polynomial approximates the function f except near its discontinuities.

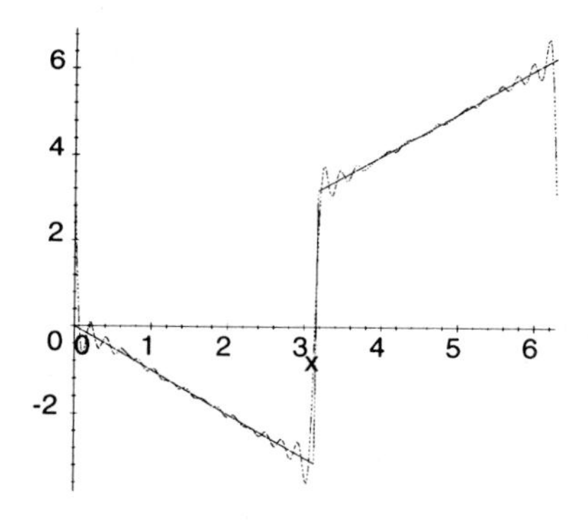

To see an animated view of the Fourier polynomials $p(n, x)$ as n ranges from 1 to 30 **click here.**

9.5.7 Another Discontinuous Function

In this subsection we shall illustrate the results of Dirichlet and Gibbs again using the function

$$f(x) = (1 + x) \left(\frac{x - \frac{\pi}{2}}{\left| x - \frac{\pi}{2} \right|} \right) \left(\frac{x - \pi}{|x - \pi|} \right) \left(\frac{x - \frac{3\pi}{2}}{\left| x - \frac{3\pi}{2} \right|} \right)$$

for $x \in [0, 2\pi)$. We observe that

$$f(x) = \begin{cases} -(1 + x) & \text{if} \quad 0 < x < \frac{\pi}{2} \\ 1 + x & \text{if} \quad \frac{\pi}{2} < x < \pi \\ -(1 + x) & \text{if} \quad \pi < x < \frac{3\pi}{2} \\ 1 + x & \text{if} \quad \frac{3\pi}{2} < x < 2\pi \end{cases}$$

And so the Fourier polynomial

$p(n,x) = \frac{1}{2\pi}\int_0^{2\pi} f(t)\,dt + \frac{1}{\pi}\sum_{j=1}^{n}\left(\left(\int_0^{2\pi} f(t)\cos jt dt\right)\cos jx + \left(\int_0^{2\pi} f(t)\sin jt dt\right)\sin jx\right)$

can be expressed in the form[21]

$p(n,x) =$

$-\frac{1}{2\pi}\int_0^{\frac{\pi}{2}} (1+t)\,dt - \frac{1}{\pi}\sum_{j=1}^{n}\left(\left(\int_0^{\frac{\pi}{2}} (1+t)\cos jt dt\right)\cos jx + \left(\int_0^{\frac{\pi}{2}} (1+t)\sin jt dt\right)\sin jx\right) +$

$\frac{1}{2\pi}\int_{\frac{\pi}{2}}^{\pi} (1+t)\,dt + \frac{1}{\pi}\sum_{j=1}^{n}\left(\left(\int_{\frac{\pi}{2}}^{\pi} (1+t)\cos jt dt\right)\cos jx + \left(\int_{\frac{\pi}{2}}^{\pi} (1+t)\sin jt dt\right)\sin jx\right) -$

$\frac{1}{2\pi}\int_{\pi}^{\frac{3\pi}{2}} (1+t)\,dt - \frac{1}{\pi}\sum_{j=1}^{n}\left(\left(\int_{\pi}^{\frac{3\pi}{2}} (1+t)\cos jt dt\right)\cos jx + \left(\int_{\pi}^{\frac{3\pi}{2}} (1+t)\sin jt dt\right)\sin jx\right) +$

$\frac{1}{2\pi}\int_{\frac{3\pi}{2}}^{2\pi} (1+t)\,dt + \frac{1}{\pi}\sum_{j=1}^{n}\left(\left(\int_{\frac{3\pi}{2}}^{2\pi} (1+t)\cos jt dt\right)\cos jx + \left(\int_{\frac{3\pi}{2}}^{2\pi} (1+t)\sin jt dt\right)\sin jx\right).$

For example,

$p(60,x) = -\frac{1}{2\pi}\int_0^{\frac{\pi}{2}} (1+t)\,dt - \frac{1}{\pi}\sum_{j=1}^{60}\left(\left(\int_0^{\frac{\pi}{2}} (1+t)\cos jt dt\right)\cos jx + \left(\int_0^{\frac{\pi}{2}} (1+t)\sin jt dt\right)\sin jx\right.$

$\frac{1}{2\pi}\int_{\frac{\pi}{2}}^{\pi} (1+t)\,dt + \frac{1}{\pi}\sum_{j=1}^{60}\left(\left(\int_{\frac{\pi}{2}}^{\pi} (1+t)\cos jt dt\right)\cos jx + \left(\int_{\frac{\pi}{2}}^{\pi} (1+t)\sin jt dt\right)\sin jx\right) -$

$\frac{1}{2\pi}\int_{\pi}^{\frac{3\pi}{2}} (1+t)\,dt - \frac{1}{\pi}\sum_{j=1}^{60}\left(\left(\int_{\pi}^{\frac{3\pi}{2}} (1+t)\cos jt dt\right)\cos jx + \left(\int_{\pi}^{\frac{3\pi}{2}} (1+t)\sin jt dt\right)\sin jx\right) +$

$\frac{1}{2\pi}\int_{\frac{3\pi}{2}}^{2\pi} (1+t)\,dt + \frac{1}{\pi}\sum_{j=1}^{60}\left(\left(\int_{\frac{3\pi}{2}}^{2\pi} (1+t)\cos jt dt\right)\cos jx + \left(\int_{\frac{3\pi}{2}}^{2\pi} (1+t)\sin jt dt\right)\sin jx\right)$

which yields[22]

$p(60,x) = 7.0\times10^{-11}\sin 17.0x + 2.0\cos x + 7.0\times10^{-10}\sin x - 7.407407\times10^{-2}\cos 27.0x +$
$1.7\times10^{-11}\sin 59.0x - 4.651163\times10^{-2}\cos 43.0x + 2.546479\times10^{-2}\cos 10.0x - 3.0\times$
$10^{-11}\sin 29.0x + 3.773585\times10^{-2}\cos 53.0x - 5.27324\sin 2.0x + 3.766981\times10^{-3}\cos 26.0x -$
$.3515493\sin 30.0x + .1176471\cos 17.0x + 4.0\times10^{-11}\sin 8.0x + 4.081633\times10^{-2}\cos 49.0x -$
$.2775389\sin 38.0x - 5.128205\times10^{-2}\cos 39.0x + .6366198\cos 2.0x - 3.0\times10^{-11}\sin 25.0x -$
$3.0\times10^{-11}\sin 31.0x + 8.732782\times10^{-4}\cos 54.0x + 2.12472\times10^{-13}\cos 40.0x + .2222222\cos 9.0x -$
$2.0\times10^{-11}\sin 55.0x + .08\cos 25.0x + 1.299224\times10^{-2}\cos 14.0x + 6.896552\times10^{-2}\cos 29.0x -$
$1.0\times10^{-11}\sin 13.0x + 7.569795\times10^{-4}\cos 58.0x + 1.018592\times10^{-3}\cos 50.0x - 4.0\times$
$10^{-12}\sin 56.0x + 2.0\times10^{-11}\sin 53.0x - .3101906\sin 34.0x + 2.091002\times10^{-13}\cos 56.0x -$
$.2511066\sin 42.0x + .1538462\cos 13.0x - 1.0\times10^{-11}\sin 23.0x + 1.76349\times10^{-3}\cos 38.0x +$
$2.074061\times10^{-13}\cos 12.0x + 2.056626\times10^{-13}\cos 24.0x - 8.695652\times10^{-2}\cos 23.0x +$
$1.988697\times10^{-13}\cos 28.0x + 2.069014\times10^{-13}\cos 36.0x - 1.0\times10^{-10}\sin 11.0x + 4.0\times$
$10^{-12}\sin 52.0x + 2.083272\times10^{-13}\cos 4.0x - .1333333\cos 15.0x - 1.054648\sin 10.0x -$
$1.4\times10^{-11}\sin 33.0x + .7853982 - .7533199\sin 14.0x - .1818182\cos 11.0x + 1.443582\times$
$10^{-3}\cos 42.0x - .4793854\sin 22.0x + 2.075377\times10^{-13}\cos 16.0x - .4056338\sin 26.0x -$
$5.714286\times10^{-2}\cos 35.0x - 2.0\times10^{-11}\sin 43.0x + 1.0\times10^{-10}\sin 5.0x - 3.636364\times$
$10^{-2}\cos 55.0x - 3.921569\times10^{-2}\cos 51.0x - 3.0\times10^{-11}\sin 21.0x + 5.26132\times10^{-3}\cos 22.0x +$
$2.829421\times10^{-3}\cos 30.0x + 2.5\times10^{-11}\sin 37.0x - .6666667\cos 3.0x + 2.202837\times$
$10^{-3}\cos 34.0x + 2.014902\times10^{-13}\cos 44.0x + 4.878049\times10^{-2}\cos 41.0x + 2.00975\times$
$10^{-13}\cos 48.0x + 1.0\times10^{-11}\sin 47.0x + 2.070443\times10^{-13}\cos 8.0x + .0952381\cos 21.0x +$
$.4\cos 5.0x + 1.909859\times10^{-13}\cos 60.0x + 5.405405\times10^{-2}\cos 37.0x + 2.010764\times$
$10^{-13}\cos 52.0x + 2.042152\times10^{-13}\cos 20.0x - 3.389831\times10^{-2}\cos 59.0x - .1953052\sin 54.0x +$
$7.859503\times10^{-3}\cos 18.0x - .1052632\cos 19.0x + 6.060606\times10^{-2}\cos 33.0x + 3.4\times$
$10^{-11}\sin 41.0x - 1.757747\sin 6.0x + 4.444444\times10^{-2}\cos 45.0x - .2109296\sin 50.0x +$

---

21 The following expression will be fully visible only in the on-screen version of this text.

22 The following expression will be fully visible only in the on-screen version of this text.

$7.073553 \times 10^{-2} \cos 6.0x - 6.451613 \times 10^{-2} \cos 31.0x + 9.0 \times 10^{-12} \sin 45.0x - 4.255319 \times 10^{-2} \cos 47.0x + 1.0 \times 10^{-11} \sin 57.0x - .5859155 \sin 18.0x - .2857143 \cos 7.0x + 1.20344 \times 10^{-3} \cos 46.0x + 3.508772 \times 10^{-2} \cos 57.0x + 5.0 \times 10^{-11} \sin 7.0x + 2.10597 \times 10^{-13} \cos 32.0x - .1818358 \sin 58.0x - .2292713 \sin 46.0x - 1.0 \times 10^{-11} \sin 35.0x.$

To illustrate the polynomial $p(50, x)$, point at the expression $f(x)$ and click on Plot 2D and Rectangular. Open the plot properties dialog box, move to the page Plot Components and set the domain interval as $[0, 6.283]$. Now highlight the expression that we obtained for $p(50, x)$ and drag it into the sketch. Open the plot properties dialog box again, move to the page Plot Components, go to Item Number 2, set the color as red. You will now see the two graphs as shown in the next figure. In this figure, the graph $y = f(x)$ is shown in blue and the graph $y = p(50, x)$ is shown in red. Notice once again how closely the Fourier polynomial approximates the given function f except near its discontinuities.

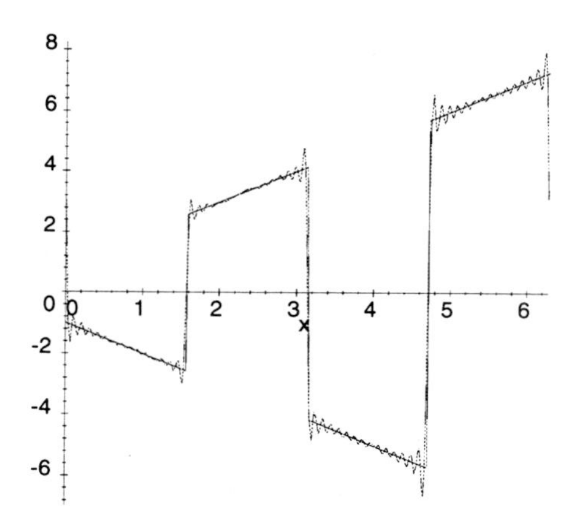

To see an animated view of the Fourier polynomials $p(n, x)$ as n ranges from 1 to 30 **click here.**

9.5.8 Exercises on Fourier Polynomials

For each of the following functions, calculate some of its Fourier polynomials and explore the way in which the sequence of Fourier polynomials converges by drawing some graphs. Finally, by calculating the integral

$$\int_0^{2\pi} (f(x) - p(n, x))^2 \, dx$$

for a variety of values of n, demonstrate that

$$\lim_{n \to \infty} \int_0^{2\pi} (f(x) - p(n, x))^2 \, dx = 0.$$

1. $f(x) = |\sin 2x|$ for $0 \le x \le 2\pi$.
2. $f(x) = x \sin(x^2 - 5x + 1)$ for $0 \le x < 2\pi$.
3. $f(x) = x \sin \frac{1}{x}$ for $x \in [-\pi, \pi) \setminus \{0\}$.
4. $f(x) = x + (1 + x)\left(\frac{x-\pi}{|x-\pi|}\right)$ for $0 \le x < 2\pi$.
5. Explore a Fourier series that has cosine terms only but converges to $\sin x$ for every $x \in [0, \pi]$. *Hint:* Define $f(x) = |\sin x|$ for $x \in [-\pi, \pi]$.
6. Explore a Fourier series that has sine terms only but converges to $\cos x$ whenever $0 < x < \pi$.

9.6 Comparing the Sums of Fourier Series

In this section we shall explore some ways in which *Scientific Notebook* can be used to determine when two trigonometric sums

$$f(x) = \sum_{n=1}^{\infty} a_n \sin nx$$

and

$$g(x) = \sum_{n=1}^{\infty} b_n \sin nx$$

will satisfy the inequality $f(x) \ge g(x)$ for every number $x \in [0, \pi]$. We begin by stating a known theorem of this type:

9.6.1 A Sufficient Condition for an Inequality

Suppose that (a_n) and (b_n) are sequences of nonnegative numbers and that

$$\sum_{n=2}^{\infty} n(a_n + b_n) < \frac{1}{\sqrt{2}}(a_1 - b_1)$$

Then the inequality

$$\sum_{n=1}^{\infty} a_n \sin nx \ge \sum_{n=1}^{\infty} b_n \sin nx$$

will hold for every number $x \in [0, \pi]$.

Although the hypotheses of this theorem are sufficient for the conclusion to hold, they are by no means necessary. We shall explore a variety of examples and demonstrate how the graphical capabilities of *Scientific Notebook* can be used to determine whether or not the conclusion of the theorem holds in each of them.

9.6.2 Some Illustrative Examples

We begin with a simple example that can be analyzed directly:

9.6.2.1 A Simple Example

In this example we take

$$f(x) = \sin x + c \sin 2x$$

for $x \in [0, \pi]$ where c is a given nonnegative number and we seek conditions on c for the inequality $f(x) \geq 0$ to hold for every $x \in [0, \pi]$. To view this question graphically we define

$$g(x, c) = \sin x + c \sin 2x$$

and we supply this definition to *Scientific Notebook* by clicking on Define and New Definition. To draw the graph we point at the expression $g(x, c)$ and click on Plot 3D. Then, to modify the graph, we open the plot properties dialog box, move to the plot components page and set the x interval $[0, 3.142]$ and c to the interval $[0, 1]$. After rotating the figure we obtain it as follows:

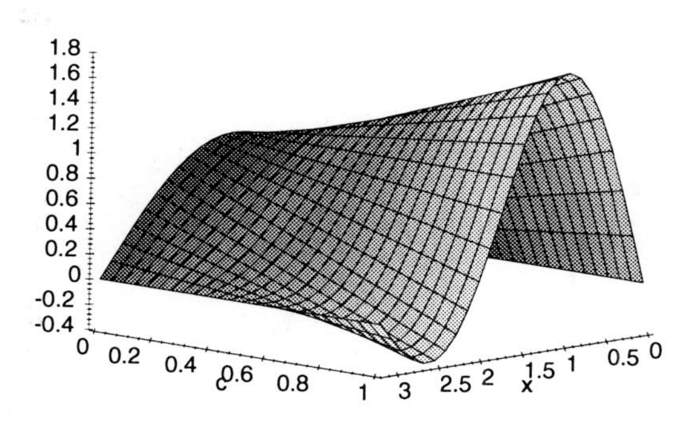

Try several other views of this figure. We can see from these figures that the value of $g(x, c)$ dips below 0 when x and c approach the upper limits of their ranges. Now, to examine the values of this function more carefully when $c = .6$, point at the expression $g(x, .6)$ and click on Plot 2D. After setting the x range to $[0, 3.142]$ you will obtain the curve shown:

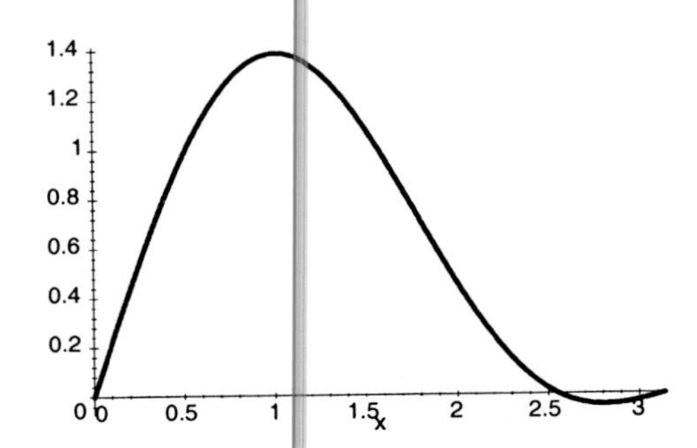

We now examine the expression $f(x)$ analytically. First we note that $f(0) = 0$ and that $f'(0) = 1 + 2c > 0$. Therefore $f(x) > 0$ whenever x is small enough and so the inequality $f(x) \geq 0$ will hold for all $x \in [0, \pi]$ as long as the equation $f(x) = 0$ has no solution between 0 and π.

Now since

$$\sin x + c \sin 2x = \sin x + 2c \sin x \cos x,$$

the equation $f(x) = 0$ will hold at a number x between 0 and π if and only if $1+2c\cos x = 0$. This condition requires that $\cos x = -\frac{1}{2c}$ and is therefore possible if and only if $c > \frac{1}{2}$. Thus if $c \leq 1$, the function f must be nonnegative. By drawing the graph of this function when $c = 0.47$ you can verify that the function is nonnegative in this case.

9.6.2.2 Another Trigonometric Polynomial

In this example we define

$$f(x) = \sin x + \frac{1}{2} \sin 2x + \frac{1}{4} \sin 4x + \frac{1}{5} \sin 5x$$

and we supply this definition to *Scientific Notebook* by clicking on Define and New Definition. If we draw the graph of this function and set its domain to $[0, \pi]$ we obtain the following figure:

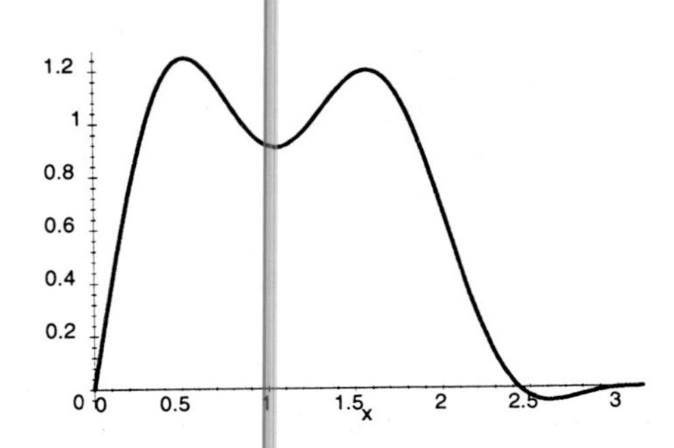

Looking at this figure we can see that the inequality $f(x) \geq 0$ does not hold for every

$x \in [0, \pi]$.

9.6.2.3 An Infinite Series

In this example we define

$$f(x) = \sum_{j=1}^{\infty} \frac{\sin jx}{\sqrt{j}}$$

for all $x \in [0, \pi]$. In order to test the inequality, we look at the partial sums of this series for some large values of n. For this purpose we define

$$f(n, x) = \sum_{j=1}^{n} \frac{\sin jx}{\sqrt{j}}$$

for each positive integer n and $x \in [0, \pi]$ and we supply this definition to *Scientific Notebook* in the usual way. We can now sketch the graphs of $f(n, x)$ for some chosen values of n. For example, the graph of $f(20, x)$ appears as follows:

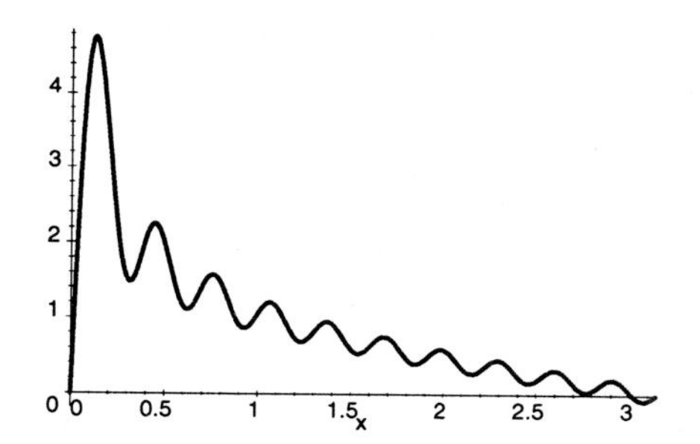

This graph does dip below the x-axis. Now draw some of these graphs for larger values of n. You will find that, as n increases, the graph looks more and more like the graph of a nonnegative function. Our final conclusion is therefore that the inequality $f(x) \geq 0$ probably does hold for all $x \in [0, \pi]$.

Another way in which you can use *Scientific Notebook* to conclude that $f(x) \geq 0$ for each x is to evaluate $f(n, x)$ for a large value of n and a value of x slightly below π. For example,

$$f(1000, 3.13) = 1.7427 \times 10^{-2}$$

which is positive.

9.6.3 Some Exercises

In each of the following examples, use *Scientific Notebook* to determine whether or not the inequality $f(x) \geq 0$ holds for every number x in the interval $[0, \pi]$.

1. For every $x \in [0, \pi]$ we define

$$f(x) = \sin x + \frac{1}{2}\sin 2x + \frac{1}{3}\sin 3x + \frac{1}{4}\sin 4x.$$

2. For every $x \in [0, \pi]$ we define

$$f(x) = \sin x + \frac{1}{2}\sin 2x + \frac{1}{4}\sin 4x + \frac{1}{8}\sin 8x.$$

3. For every $x \in [0, \pi]$ we define

$$f(x) = \sum_{j=1}^{\infty} \frac{\sin jx}{j}.$$

See Subsection 10.3.5 for a discussion of this function.

4. For every $x \in [0, \pi]$ we define

$$f(x) = \sum_{j=1}^{\infty} \frac{\sin jx}{\sqrt[3]{j}}.$$

5. For every $x \in [0, \pi]$ we define

$$f(x) = \sum_{j=1}^{\infty} \frac{\sin jx}{j^2}.$$

6. For every $x \in [0, \pi]$ we define

$$f(x) = \sum_{n=1}^{\infty} \left(\frac{2}{n^3} - \frac{1}{n^2}\right) \sin nx.$$

7. For every $x \in [0, \pi]$ we define

$$f(x) = \sum_{n=1}^{\infty} \left(\frac{6}{n^4} - \frac{2}{n^3}\right) \sin nx.$$

9.6.4 A Multivariable Example

In this final example we show how *Scientific Notebook* can be used to compare the functions f and g defined by the equations

$$f(x, y) = \sum_{j=1}^{\infty}\sum_{k=1}^{\infty} \frac{\cos jx \cos ky}{j^2 k^2}$$

and

$$g(x,y) = \sum_{j=1}^{\infty}\sum_{k=1}^{\infty} \frac{4\cos jx \cos ky}{j^3k^3}$$

for all points $(x,y) \in \left[0, \frac{\pi}{2}\right] \times \left[0, \frac{\pi}{2}\right]$. For all positive integers m and n we define

$$f(m,n,x,y) = \sum_{j=1}^{m}\sum_{k=1}^{n} \frac{\cos jx \cos ky}{j^2k^2}$$

and

$$g(m,n,x,y) = \sum_{j=1}^{m}\sum_{k=1}^{n} \frac{4\cos jx \cos ky}{j^3k^3}$$

and we supply these definitions to *Scientific Notebook* in the usual way. If we draw the graphs $z = f(10,10,x,y)$ and $z = g(10,10,x,y)$ together and rotate the figure accordingly we obtain the following figure:

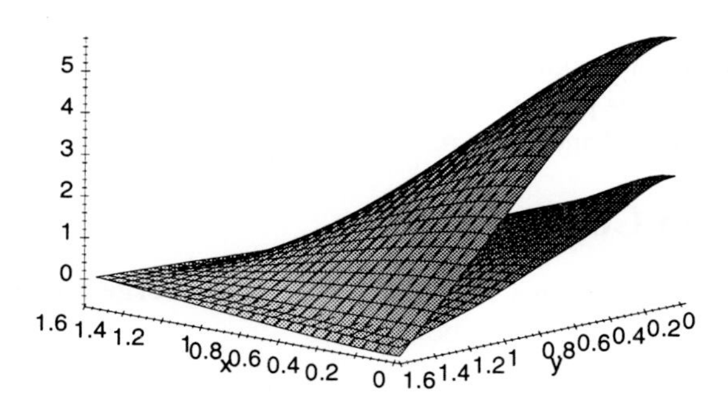

At first it appears that the one surface lies totally above the other. However, if we take a closer look at the part of this figure in which y varies from 1.3 to 1.57 and x varies from 0 to 1.3, we shall see that the two surfaces actually do cross each other.

Chapter 10 Sequences of Functions

10.1 Introduction

In this section we shall show how the graphical tools provided by *Scientific Notebook* can help us to explore the way in which a given sequence (f_n) of functions may converge to a function f as $n \to \infty$. Some of the basic types of convergence are as follows:

10.1.1 Pointwise Convergence

A sequence (f_n) of functions from a set S to $\mathbf{R}$ is said to **converge pointwise** to a function $f : S \to \mathbf{R}$ if for every $x \in S$ we have $f_n(x) \to f(x)$ as $n \to \infty$.

10.1.2 Bounded Convergence

A sequence (f_n) of functions from a set S to $\mathbf{R}$ is said to **converge boundedly** to a function $f : S \to \mathbf{R}$ if (f_n) converges pointwise to f and it is possible to find a number K such that $|f_n(x)| \leq K$ for every $x \in S$ and every n.

10.1.3 Uniform Convergence

A sequence (f_n) of functions from a set S to $\mathbf{R}$ is said to **converge uniformly** to a function $f : S \to \mathbf{R}$ if $\sup |f_n - f| \to 0$ as $n \to \infty$.

10.2 The Basic Facts about Sequences of Functions

10.2.1 Relationship between the Types of Convergence

1. If $f_n \to f$ uniformly then $f_n \to f$ pointwise.
2. If $f_n \to f$ uniformly then $f_n \to f$ boundedly if and only if the function f is bounded.

10.2.2 Uniform Convergence and Continuity

One of the key facts about uniform convergence is that the limit of a uniformly convergent sequence of continuous functions is continuous. This theorem can be stated more precisely as follows:

If $f_n \to f$ uniformly on a set S of numbers and $x \in S$ and every one of the functions f_n is continuous at the number x, then the function f is also continuous at the number x.

10.2.3 Bounded Convergence and Integration

The principal theorem that deals with the interchange of limits and Riemann integrals is the Arzela **bounded convergence theorem.** This theorem may be stated as follows:[23]

If (f_n) is a sequence of Riemann integrable functions on an interval $[a, b]$ that converges boundedly to function f then the sequence of integrals $\int_a^b f_n$ must converge. Furthermore, in the event that the limit function f is also Riemann integrable we have

$$\int_a^b f_n \to \int_a^b f$$

as $n \to \infty$.

10.3 Some Examples of Sequences of Functions

10.3.1 A Sequence with a Discontinuous Limit

For every natural number n and every real number x we define

$$f_n(x) = \frac{x^{2n}}{1 + x^{2n}}.$$

Each of the functions f_n is a rational function with a never vanishing denominator and is therefore continuous at every real number.

In the event that $|x| < 1$ it follows from the fact that $x^{2n} \to 0$ as $n \to \infty$ that $f_n(x) \to 0$ as $n \to \infty$. Whenever $|x| > 1$ we see that

$$f_n(x) = \frac{x^{2n}}{1 + x^{2n}} = \frac{1}{\frac{1}{x^{2n}} + 1} \to 1$$

as $n \to \infty$. Therefore the sequence (f_n) converges pointwise on $\mathbf{R}$ to the function f defined

---

[23] A complete discussion of this theorem and an elementary proof is provided in the book, *An Introduction to Mathematical Analysis* by Jonathan Lewin and Myrtle Lewin (McGraw Hill, 1988).

by

$$f(x) = \begin{cases} 0 & \text{if} \quad |x| < 1 \\ \frac{1}{2} & \text{if} \quad |x| = 1 \\ 1 & \text{if} \quad |x| > 1 \end{cases}.$$

Since the limit function f fails to be continuous at -1 and at 1, the convergence of the sequence (f_n) cannot be uniform.

Now, in order to use *Scientific Notebook* to explore the graphs of the functions f_n we shall change our notation and write $f(n,x)$ instead of $f_n(x)$. In other words, we define

$$f(n,x) = \frac{x^{2n}}{1+x^{2n}}$$

for every natural number n and every real number x and we inform *Scientific Notebook* of this definition by pointing at it and clicking on Define and New Definition.

To draw the graph of the function f_n for any specified value of n we simply point at the expression $f(n,x)$ and click on Plot 2D and Rectangular. For example, if we apply this procedure to the function f_5 then we obtain the following graph:

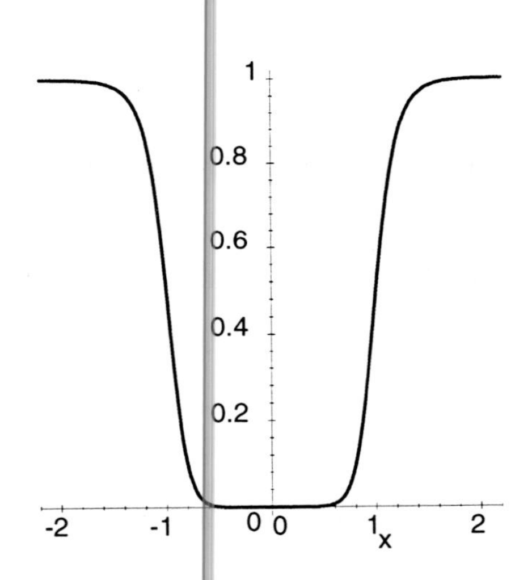

In order to explore this sequence further, follow the following procedure:

- Point at the expression $f(1,x)$ and click on Plot 2D and Rectangular.
- Zoom in a little so that the domain of your plot is approximately $[-2,2]$.
- Highlight the expression $f(2,x)$ and drag this expression into your plot.
- Highlight the expression $f(n,x)$ for several other values of n of your choosing and drag each of these into your plot.
- Experiment with different colors for your different plots to make it easy to distinguish between them. You may find that your figure is easier to read if you set each plot thickness as Medium instead of Thin.

The graphs you will obtain in this way will help you to visualize the fact that the limit function of this sequence (f_n) is discontinuous at the number 1. To see an animated view of these graphs as n runs from 1 to 20, **click here**.

10.3.2 A Uniformly Convergent Sequence

For every natural number n and every number $x \in [0,1]$ we define

$$f_n(x) = nx \exp\left(-n^2 x\right)$$

Each of the functions f_n is continuous on the interval $[0,1]$. Now given any number $x \in [0,1]$ it is easy to see that

$$\lim_{n \to \infty} nx \exp\left(-n^2 x\right) = 0$$

and so we deduce that the sequence (f_n) converges pointwise to the constant function 0. To show that the sequence actually converges uniformly to 0 we observe that, for each n, the function f_n must take its maximum value either at 0 or at 1 or at a value of x for which $f_n'(x) = 0$. Since

$$f_n'(x) = n \exp\left(-n^2 x\right) - n^3 x \exp\left(-n^2 x\right)$$

we see that $f_n'(x) = 0$ when $x = 1/n^2$. We deduce that if n is sufficiently large then

$$\sup f_n = \max f_n = f_n\left(\frac{1}{n^2}\right) = \frac{1}{ne}$$

and the uniform convergence of (f_n) to 0 follows since the latter expression approaches 0 as $n \to \infty$.

It is worth noticing that

$$\lim_{n \to \infty} \int_0^1 f_n(x)\,dx = \lim_{n \to \infty} \frac{1 - e^{-n^2} n^2 - e^{-n^2}}{n^3} = 0 = \int_0^1 \lim_{n \to \infty} f_n(x)\,dx.$$

We shall now use *Scientific Notebook* to explore the graphs of the functions f_n. As before, we shall change our notation and write $f(n,x)$ instead of $f_n(x)$. In other words, we define

$$f(n,x) = nx \exp\left(-n^2 x\right)$$

for every natural number n and every number $x \in [0,1]$ and we inform *Scientific Notebook* of this definition by pointing at it and clicking on Define and New Definition.

To draw the graph of the function f_n for any specified value of n we simply point at the expression $f(n,x)$ and click on Plot 2D and Rectangular. For example, if we apply this procedure to the function f_3 then we obtain the graph shown.

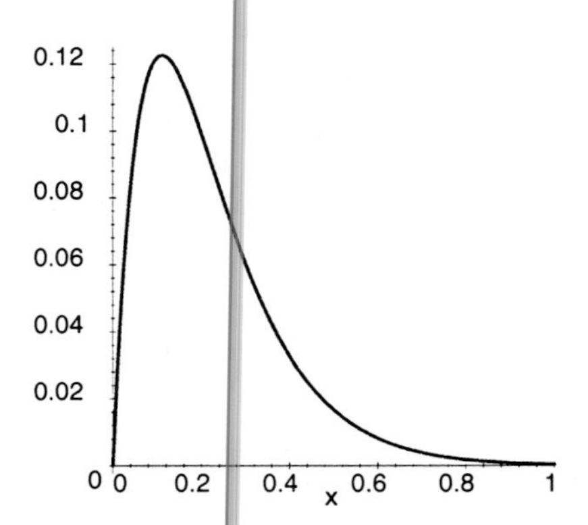

In order to explore this sequence further, follow the following procedure:

- Point at the expression $f(1,x)$ and click on Plot 2D and Rectangular.
- Open the plot properties dialog box, move to Plot Components and set the domain interval as $[0,1]$.
- Highlight the expression $f(2,x)$ and drag this expression into your plot.
- Highlight the expression $f(n,x)$ for several other values of n of your choosing and drag each of these into your plot.
- Experiment with different colors for your different plots to make it easy to distinguish between them. You may find that your figure is easier to read if you set each plot thickness as Medium instead of Thin.

The graphs you will obtain in this way will help you to visualize the fact that the sequence (f_n) converges uniformly to 0. As you will see, although each graph has a spike, these spikes move to the left and become lower and lower as n increases. This illustrates the fact that $\sup f_n \to 0$ as $n \to \infty$. To see an animated view of these graphs as n runs from 1 to 20, **click here**.

10.3.3 A Boundedly Convergent Sequence

For every natural number n and every number $x \in [0,1]$ we define

$$f_n(x) = nx \exp(-nx)$$

Each of the functions f_n is continuous on the interval $[0,1]$. Now given any number $x \in [0,1]$ it is easy to see that

$$\lim_{n\to\infty} nx \exp(-nx) = 0$$

and so we deduce that the sequence (f_n) converges pointwise to the constant function 0.

As in the preceding example we observe that each function f_n must take its maximum value either at 0 or at 1 or at a value of x for which $f_n'(x) = 0$. The latter equation holds when

$$ne^{-nx} - n^2xe^{-nx} = 0$$

which requires that $x = \frac{1}{n}$. Thus each function f_n has a maximum value of $f_n(1/n) = 1/e$ and we conclude that the sequence (f_n) does not converge uniformly to the constant function 0. However, this sequence does converge boundedly. Notice that

$$\lim_{n\to\infty} \int_0^1 f_n(x)\,dx = \lim_{n\to\infty} \frac{1 - e^{-n} - e^{-n}n}{n} = 0 = \int_0^1 \lim_{n\to\infty} f_n(x)\,dx.$$

We shall now use *Scientific Notebook* to explore the graphs of the functions f_n. As before, we shall change our notation and write $f(n, x)$ instead of $f_n(x)$. In other words, we define

$$f(n, x) = nx\exp(-nx)$$

for every natural number n and every number $x \in [0, 1]$ and we inform *Scientific Notebook* of this definition by pointing at it and clicking on Define and New Definition.

To draw the graph of the function f_n for any specified value of n we simply point at the expression $f(n, x)$ and click on Plot 2D and Rectangular. For example, if we apply this procedure to the function f_7 then we obtain the graph shown:

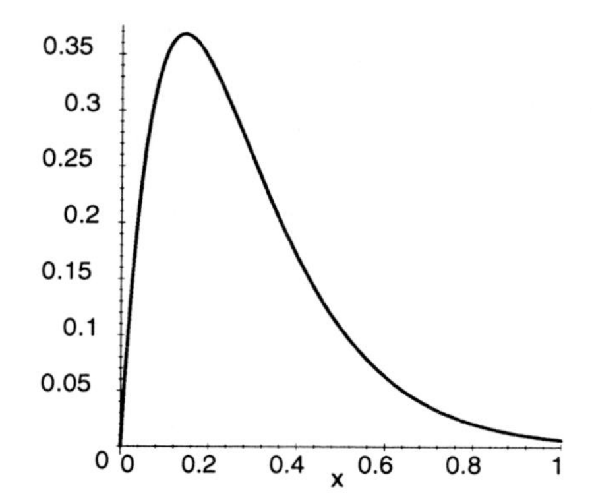

In order to explore this sequence further, follow the following procedure:

- Point at the expression $f(1, x)$ and click on Plot 2D and Rectangular.
- Open the plot properties dialog box, move to Plot Components and set the domain interval as $[0, 1]$.
- Highlight the expression $f(2, x)$ and drag this expression into your plot.
- Highlight the expression $f(n, x)$ for several other values of n of your choosing and drag each of these into your plot.
- Experiment with different colors for your different plots to make it easy to distinguish between them. You may find that your figure is easier to read if you set each plot thickness as Medium instead of Thin.

The graphs you will obtain in this way will help you to visualize the fact that the sequence (f_n) converges boundedly but not uniformly to 0. As you will see, each graph has a spike of height 0.35 and these spikes move to the left as n increases. To see an animated view of these graphs as n takes some assorted values from 1 to 100, **click here**.

One more experiment that you may like to try is to look at the graphs of the functions f_{10^6} and f_{10^7} on the domain $\left[0, 10^{-6}\right]$.

10.3.4 A Sequence that Converges Pointwise but Not Boundedly

For every natural number n and every number $x \in [0,1]$ we define

$$f_n(x) = n^2 x \exp(-nx)$$

Each of the functions f_n is continuous on the interval $[0,1]$. Now given any number $x \in [0,1]$ it is easy to see that

$$\lim_{n \to \infty} n^2 x \exp(-nx) = 0$$

and so we deduce that the sequence (f_n) converges pointwise to the constant function 0. Arguing as we did in the preceding examples we can see that the function f_n has a maximum value at $\frac{1}{n}$ and that this maximum value is $\frac{n}{e}$. Therefore $\sup f_n \to \infty$ as $n \to \infty$ and we conclude that the sequence (f_n) does not converge boundedly.

Notice that

$$\lim_{n \to \infty} \int_0^1 f_n(x)\, dx = 1 - \frac{8}{e^7} \neq \int_0^1 \lim_{n \to \infty} f_n(x)\, dx.$$

We shall now use *Scientific Notebook* to explore the graphs of the functions f_n. As before, we shall change our notation and write $f(n,x)$ instead of $f_n(x)$. In other words, we define

$$f(n,x) = n^2 x \exp(-nx)$$

for every natural number n and every number $x \in [0,1]$ and we inform *Scientific Notebook* of this definition by pointing at it and clicking on Define and New Definition.

To draw the graph of the function f_n for any specified value of n we simply point at the expression $f(n,x)$ and click on Plot 2D and Rectangular. For example, if we apply this procedure to the function f_7 then we obtain the graph shown:

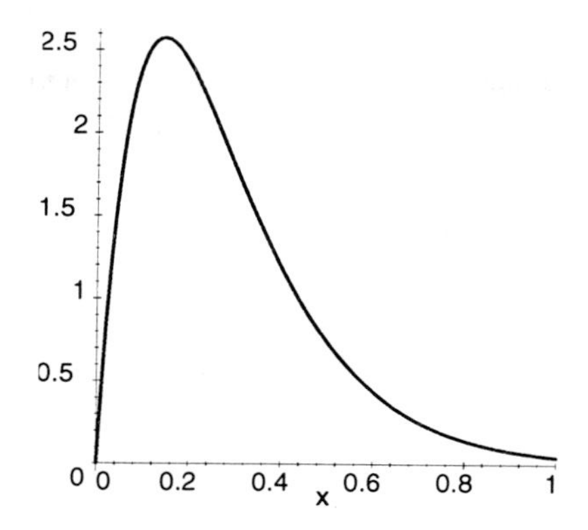

In order to explore this sequence further, follow the following procedure:

- Point at the expression $f(1,x)$ and click on Plot 2D and Rectangular.
- Open the plot properties dialog box, move to Plot Components and set the domain interval as $[0,1]$.
- Move to the Axes & View page in the plot properties dialog box and deselect the default button for the View Intervals. Set the view intervals as shown with x running from 1 to 2 and y running from 0 to 11.25. Click on OK.

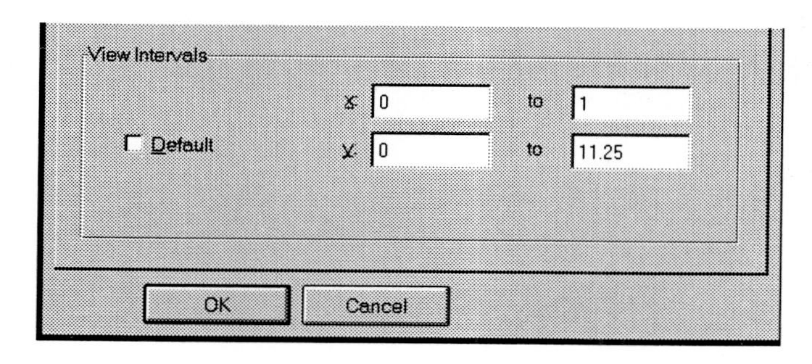

- Highlight the expression $f(2,x)$ and drag this expression into your plot.
- Highlight the expression $f(n,x)$ for several other values of n of your choosing and drag each of these into your plot.
- Experiment with different colors for your different plots to make it easy to distinguish between them. You may find that your figure is easier to read if you set each plot thickness as Medium instead of Thin.

The graphs you will obtain in this way will help you to visualize the fact that the sequence (f_n) converges pointwise but not boundedly to 0. As you will see, each graph has a spike and these spikes move to the left and become higher and higher as n increases. This illustrates the fact that $\sup f_n \to \infty$ as $n \to \infty$. To see an animated view of these graphs as n takes some assorted values from 1 to 30, **click here**.

10.3.5 A Sequence of Fourier Polynomials

In this subsection we consider the sequence (p_n) of trigonometric polynomials where for each positive integer n and each number $x \in [0, 2\pi]$ we define

$$p_n(x) = \sum_{j=1}^{n} \frac{\sin jx}{j}.$$

It can be shown that whenever $0 < x < 2\pi$ we have

$$p_n(x) \to \frac{\pi - x}{2}$$

as $n \to \infty$ and it is clear that $p_n(0) = p_n(2\pi) = 0$ for every n. Thus the sequence (p_n) converges pointwise on $[0, 2\pi]$ to the function f which is defined as

$$f(x) = \begin{cases} 0 & \text{if } x = 0 \\ \frac{\pi - x}{2} & \text{if } 0 < x < 2\pi \\ 0 & \text{if } x = 2\pi \end{cases}.$$

In other words, for every number $x \in [0, 2\pi]$ we have

$$f(x) = \sum_{n=1}^{\infty} \frac{\sin nx}{n}.$$

The graph of this function f is illustrated in the following figure:

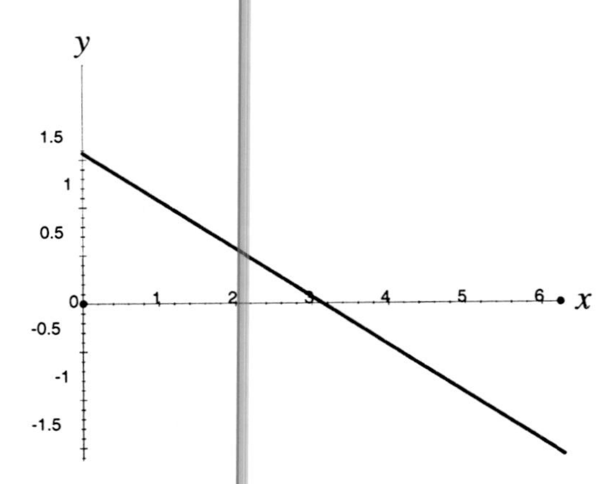

Since f is discontinuous both at 0 and at 2π we know that the convergence of the sequence (p_n) is not uniform. Since, for any given n, the quantity $p_n(x)$ can be made as small as we like by making x close enough to zero, it may be a little surprising that the limit of this sequence of trigonometric polynomials is discontinuous at 0. With the help of *Scientific Notebook* we can improve our perception of this phenomenon. First we shall make a minor change in notation and agree to write $p(n,x)$ instead of $p_n(x)$. We point at the equation

$$p(n,x)=\sum_{j=1}^{n}\frac{\sin jx}{j}$$

and click on Define and New Definition. Point at the expression $p(n,x)$ for some values of n of your choosing and click on Plot 2D, setting the plot domain as $[0, 6.283]$. For example, the graph of $p(10,x)$ comes out as follows:

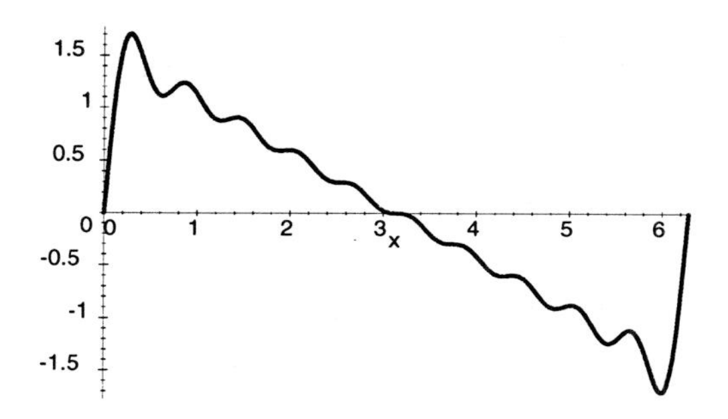

By drawing several more graphs of this type you will see that this sequence of trigonometric polynomials displays the same behavior that we saw in Subsection 10.3.3. To see an animated view of these graphs, **click here**.

Another way to appreciate the fact that the limit of this sequence of trigonometric polynomials is not continuous is to observe that for each n we have

$$p_n'(x)=\sum_{j=1}^{n}\cos jx=\frac{\sin\left(n+\frac{1}{2}\right)x-\sin\frac{x}{2}}{2\sin\frac{x}{2}}.$$

Therefore $p_n'(x)=0$ when

$$\sin\left(n+\frac{1}{2}\right)x=\sin\frac{x}{2},$$

and the smallest positive root of this equation is $x=\frac{\pi}{n+1}$. With this in mind, we define $g(n)=p\left(n,\frac{\pi}{n+1}\right)$ and we supply this definition to *Scientific Notebook* by clicking on Define and New Definition. Now point at the expression $g(n)$ for several chosen values of n and click on Evaluate Numerically. For example, $g(10000)=1.8518$ and so $\lim_{n\to\infty}g(n)$ is approximately 1.8518.

10.3.6 Exercise

Given that p is a real number and that

$$f_n(x) = \frac{n^p x^2}{(1 + n^2 x^3)^3}$$

for every natural number n and every $x \in [0, 1]$, investigate the behavior of the sequence (f_n) as $n \to \infty$ for a variety of values of $p \leq 6$. Determine for which values of p the sequence (f_n) converges uniformly, for which values of p the sequence (f_n) converges boundedly but not uniformly and for which values of p the sequence (f_n) converges pointwise but not boundedly. In each case, investigate the behavior of the integral $\int_0^1 f_n$ as $n \to \infty$.

Use the graphical features of *Scientific Notebook* to illustrate the results you have obtained.

Chapter 11
Fixed Point Theorems

In this Chapter we shall demonstrate how *Scientific WorkPlace* can be used to find the fixed points of a given function numerically. We also give a brief discussion of two important theorems that assert the existence of fixed points. These are the theorem on fixed points of contractions and the famous Brouwer fixed point theorem.

11.1 Introduction to Fixed Points

11.1.1 Definition of a Fixed Point

If f is a function from a set A into a set B then a point a in the set A is said to be a **fixed point** of the function f if $f(a) = a$.

Thus, if A is a set of real numbers then a number a is a fixed point of f when the graph $y = f(x)$ meets the line $y = x$ at $x = a$.

11.1.2 A Function with Two Fixed Points

If $f(x) = 1 + 2\ln x$ for every positive number x then it is clear from the following figure that the graph of f meets the line $y = x$ at exactly two points. One of these points is the point $(1, 1)$ and so the number 1 is a fixed point of the function f. The other fixed point lies between 3 and 4. To find it more accurately, one may solve the equation

$$1 + 2\ln x - x = 0$$

numerically or one may zoom into the figure.

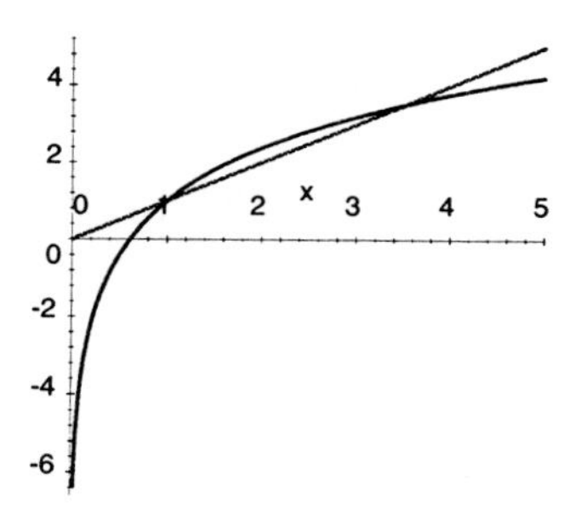

11.1.3 Some Exercises on Fixed Points

1. Given that $f(x) = 1 + 2\log x$ for $x > 0$, point at the equation $1 + 2\ln x - x = 0$ and click on Solve and Numeric to find approximations to the two fixed points of the function f.

2. Experiment with the graphs of functions f of the form

$$f(x) = a + b\log cx$$

where a, b and c are constants and determine for which choices of these constants the function f will have two different fixed points, one fixed point or no fixed point at all. Why can't a function of this type have more than two fixed points? You can find an answer to the latter question in Note 3 in Appendix A.

3. (a) By pointing at the system of equations

$$\begin{aligned} \left|\frac{1}{2} - x\right| |\cos y| &= x \\ \left|\frac{1}{2} - y\right| |\cos x| &= y \end{aligned}$$

and clicking on Solve and Numeric, find an approximation to a fixed point of the function f given by

$$f(x, y) = \left(\left|\frac{1}{2} - x\right| |\cos y|, \left|\frac{1}{2} - y\right| |\cos x|\right)$$

for all points $(x, y) \in \mathbf{R}^2$.

(b) Given that S is the triangle with vertices $(0,0)$, $(1,0)$ and $(0,1)$, explain why the function f is a continuous function from S to S and observe that the fixed point of f that you found above belongs to S.

4. Repeat the preceding exercise for the function f given by

$$f(x, y) = \left(\left|\frac{1}{2} - x\right| |\cos 5y|, \left|\frac{1}{2} - y\right| |\cos x|\right).$$

Then experiment with some functions f of the form

$$f(x, y) = \left(\left|\frac{1}{2} - x\right| |\cos my|, \left|\frac{1}{2} - y\right| |\cos nx|\right)$$

for various integers m and n.

5. By pointing at the system of equations

$$\begin{aligned} y\cos(x + y) &= x \\ (x + y)\sin z &= y \\ x\cos(y + z) &= z \end{aligned}$$

and clicking on Solve and Numeric, find an approximation to a fixed point of the function f given by

$$f(x,y,z) = (y\cos(x+y), (x+y)\sin z, x\cos(y+z))$$

whenever $(x,y,z) \in \mathbf{R}^3$.

6. Given that

$$f(x,y) = \left(\left|\frac{1}{2} - x\right| \left|\cos\left(2y + \frac{\pi}{3}\right)\right|, \left|\frac{1}{2} - y\right| |\cos x|\right)$$

for every point (x,y) in the triangle S with vertices $(0,0)$, $(1,0)$ and $(0,1)$, explain why f is a continuous function from S to S. What happens when you try to repeat the method of the preceding exercises to find a fixed point of f? Do you think that this function f has a fixed point?

11.2 Fixed Points of Contractions

11.2.1 Definition of a Contraction

If S is a set of real numbers than a function $f : S \to \mathbf{R}$ is said to be a **contraction** on S if it is possible to find a number $\delta < 1$ such that the inequality

$$|f(x) - f(t)| \leq \delta |x - t|$$

holds for all numbers x and t in the set S.

More generally, if S is a subset of the Euclidean space $\mathbf{R}^n$ then a function $f : S \to \mathbf{R}^n$ is said to be a **contraction** on S if it is possible to find a number $\delta < 1$ such that the inequality

$$\|f(x) - f(t)\| \leq \delta \|x - t\|$$

holds for all points x and t in the set S.

11.2.2 Some Special Cases of Contractions

11.2.2.1 Contractions on an Interval

Suppose that f is a continuous function from an interval $[a,b]$ to $[a,b]$ and that $\delta < 1$ and that for every number $x \in (a,b)$ we have $|f'(x)| \leq \delta$. Using the mean value theorem, one may see easily that

$$|f(x) - f(t)| \leq \delta |x - t|$$

for all numbers x and t in the interval $[a,b]$ and so f must be a contraction on $[a,b]$.

11.2.2.2 Contractions on Convex Sets

Suppose that f is a continuous function from a convex subset S of $\mathbf{R}^n$ into S and that for every point x in the interior of x we have $\|f'(x)\| \leq \delta$. Then the inequality

$$\|f(x) - f(t)\| \leq \delta \|x - t\|$$

holds for all points x and t in S and so f is a contraction on S.

Note that if a point x in $\mathbf{R}^n$ is written in the form

$$x = \begin{bmatrix} x_1 \\ x_2 \\ \vdots \\ x_n \end{bmatrix}$$

and if f is written in the form $(f_1, f_2, \cdots, f_n)$ then $\|f'(x)\|$ is the norm of the $n \times n$ matrix

$$\left[\frac{\partial f_i}{\partial x_j}\right].$$

Thus, a sufficient condition for f to be a contraction on S is that for some number $\delta < 1$ we have

$$\sum_{i=1}^{n}\sum_{j=1}^{n}\left(\frac{\partial f_i}{\partial x_j}\right)^2 \leq \delta$$

at every point of S.

11.2.3 Contractions Can Have at Most One Fixed Point

Suppose that f is a contraction on a subset S of $\mathbf{R}^n$. Then the function f can have at most one fixed point in S.

Proof. Choose a number $\delta < 1$ such that the inequality

$$\|f(x) - f(t)\| \leq \delta \|x - t\|$$

holds for all x and t in S. Now suppose that x_1 and x_2 are both fixed points of f. Since

$$\|f(x_1) - f(x_2)\| \leq \delta \|x_1 - x_2\|$$

and since $f(x_1) = x_1$ and $f(x_2) = x_2$ we have $\|x_1 - x_2\| \leq \delta \|x_1 - x_2\|$. Therefore $\|x_1 - x_2\| = 0$ and so $x_1 = x_2$.

11.2.4 Existence of Fixed Points of Contractions

Suppose that f is a contraction on a subset S of $\mathbf{R}^n$ and that $f : S \to S$. For each positive integer n we shall write f^n for the composition of the function f with itself n times.

1. *Given any point $t \in S$, the sequence $(f^n(t))$ converges to a point in $\mathbf{R}^n$. In the event that the set S is closed we have $\lim_{n\to\infty} f^n(t) \in S$.*
2. *Given any point $t \in S$, if $\lim_{n\to\infty} f^n(t) \in S$ then this limit is a fixed point of f.*
3. *If the function f has a fixed point x then for every point $t \in S$ we have*

$$\lim_{n\to\infty} f^n(t) = x.$$

4. *In the event that the set S is closed then f must have a fixed point in S.*

11.2.5 An Error Estimate

Suppose that f is a contraction on a subset S of $\mathbf{R}^n$ and that $f: S \to S$. Suppose that $\delta < 1$ and that the inequality

$$\|f(x) - f(t)\| \le \delta \|x - t\|$$

holds for all points x and t in S and suppose finally that whenever x and t are points of S, the expression $\|x - t\|$ does not exceed a constant K. As we have said, if x is the fixed point of this function f then given any $t \in S$ we have

$$\lim_{n\to\infty} f^n(t) = x.$$

Now given any n, we have

$$\|f^n(t) - x\| = \|f^n(t) - f^n(x)\| \le \delta^n \|x - t\| \le K\delta^n.$$

Thus, to find an estimate to this fixed point x within a given specified tolerance ε all we have to do is work out the expression $f^n(t)$ for a value of n large enough to ensure that $K\delta^n < \varepsilon$.

11.2.6 Example of a Contraction

We consider the function f defined by

$$f(x) = \pi + \frac{\cos x}{2}$$

for every $x \in [0, 2\pi]$. Observe that $f: [0, 2\pi] \to [0, 2\pi]$. Now since

$$|f'(x)| = \left|\frac{\sin x}{2}\right| \le \frac{1}{2}$$

for each x we know that f is a contraction on the interval $[0, 2\pi]$ and therefore f must have a unique fixed point in this interval. To find an approximation to this fixed point we select any number $t \in [0, 2\pi]$ and evaluate the limit

$$\lim_{n\to\infty} f^n(t)$$

numerically. To accomplish this we point at the definition of the function f and click on Define and New Definition. Then we obtain the iterates of the function f as

$$\begin{array}{c} 0 \\ 3.6416 \\ 2.7028 \\ 2.689 \\ 2.6919 \\ 2.6913 \\ 2.6914 \\ 2.6914 \\ 2.6914 \\ 2.6914 \\ 2.6914 \end{array}$$

Finally we observe that in order to make an expression $f^n(t)$ approximate the fixed point of f within a given tolerance ε we can select n in such a way that

$$\left(\frac{1}{2}\right)^n (2\pi) < \varepsilon.$$

For example, if we want $f^n(t)$ to approximate the fixed point to five decimal places it is sufficient to make

$$\left(\frac{1}{2}\right)^n (2\pi) < 0.00001.$$

This inequality says that

$$n > \frac{\ln\left((2\pi)\left(10^5\right)\right)}{\ln 2} = 19.261$$

and so we know that the approximation $f^n(t)$ is accurate to five decimal places whenever $n \geq 20$.

11.2.7 A Contraction Function of Two Variables

$$f(x,y) = \left(1 + \frac{1}{3}\cos xy, 1 + \frac{1}{3}\cos(x+y)\right)$$

for $0 \leq x \leq 2$ and $0 \leq y \leq 2$. We write the coordinates of $f(x,y)$ as $f_1(x,y)$ and $f_2(x,y)$ for each point (x,y) in the square $[0,2] \times [0,2]$. We observe that

$$\begin{bmatrix} \frac{\partial f_1}{\partial x} & \frac{\partial f_1}{\partial y} \\ \frac{\partial f_2}{\partial x} & \frac{\partial f_2}{\partial y} \end{bmatrix} = \begin{bmatrix} -\frac{y}{3}\sin xy & -\frac{x}{3}\sin xy \\ -\frac{1}{3}\sin(x+y) & -\frac{1}{3}\sin(x+y) \end{bmatrix}.$$

Now point at the expression

$$\left\| \begin{bmatrix} -\frac{y}{3}\sin xy & -\frac{x}{3}\sin xy \\ -\frac{1}{3}\sin(x+y) & -\frac{1}{3}\sin(x+y) \end{bmatrix} \right\|$$

and click on Evaluate. The norm is shown as the larger of two identical expressions and is therefore equal to either one of them. To obtain an upper bound for the norm of the matrix

$$\begin{bmatrix} \frac{\partial f_1}{\partial x} & \frac{\partial f_1}{\partial y} \\ \frac{\partial f_2}{\partial x} & \frac{\partial f_2}{\partial y} \end{bmatrix}$$

as the point (x, y) varies through the square $[0, 2] \times [0, 2]$ we need to maximize the function g defined by the equation[24]

$$g(x,y) = \frac{1}{6}\sqrt{\left(2\left(\sin^2 xy\right)y^2 + 2\left(\sin^2 xy\right)x^2 + 4\sin^2(x+y) + 2\sqrt{\left(\left(\sin^4 xy\right)y^4 + 2\left(\sin^4 xy\right)y^2x^2 + \right.}\right.}$$

Point at this equation and click on Define and New Definition and then click on Plot 3D and Rectangular. Open the plot properties dialog box and set the x and y intervals to be $[0, 2]$. Then drag the expression 1 into the sketch in order to add the plot $z = 1$ to what we already have. After you have rotated the plot suitably it will appear as follows,

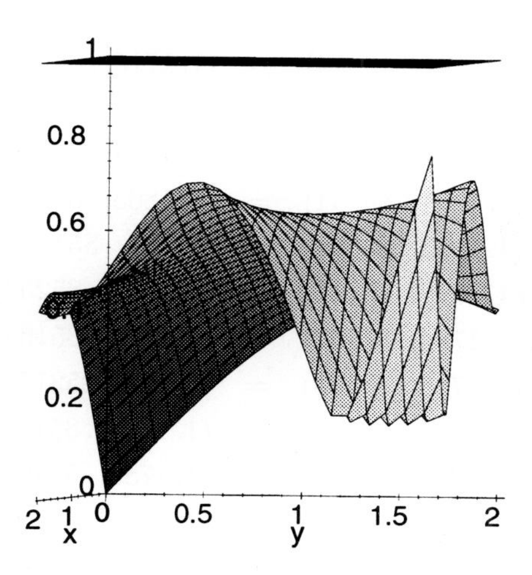

and we see from this figure that the maximum value of the function g is less than 1. We can therefore deduce from Subsection 11.2.2 that the function f is a contraction.

[24] The following expression will be fully visible only in the on-screen version of this text.

To show that the fixed point of this function is approximately $(1.1789, .852)$, point at the system of equations

$$\begin{aligned} 1+\frac{1}{3}\cos xy &= x \\ 1+\frac{1}{3}\cos(x+y) &= y \end{aligned}$$

and click on Solve and Numeric. Finally, we shall approximate this fixed point by iterating the function f starting at an arbitrary point in the square $[0,2]\times[0,2]$. Unfortunately, *Scientific Notebook* does not provide automatic iteration of a function of two variables but we can still iterate the function by hand. We see that

$$\begin{aligned}
f(0,0) &= (1.3333, 1.3333) \\
f(1.3333, 1.3333) &= (.93153, .70357) \\
f(.93153, .70357) &= (1.2643, .97858) \\
f(1.2643, .97858) &= (1.1091, .79246) \\
f(1.1091, .79246) &= (1.2127, .89174) \\
f(1.1091, .79246) &= (1.2127, .89174) \\
f(1.2127, .89174) &= (1.1567, .83044) \\
f(1.1567, .83044) &= (1.191, .86519) \\
f(1.191, .86519) &= (1.1715, .84448) \\
f(1.1715, .84448) &= (1.1831, .85646) \\
f(1.1831, .85646) &= (1.1764, .84941) \\
f(1.1764, .84941) &= (1.1803, .85351) \\
f(1.1803, .85351) &= (1.178, .85112) \\
f(1.178, .85112) &= (1.1794, .85252) \\
f(1.1794, .85252) &= (1.1786, .85168) \\
f(1.1786, .85168) &= (1.179, .85217) \\
f(1.179, .85217) &= (1.1788, .85191) \\
f(1.1788, .85191) &= (1.1789, .85204) \\
f(1.1789, .85204) &= (1.1788, .85197) \\
f(1.1788, .85197) &= (1.1789, .85203) \\
f(1.1789, .85203) &= (1.1788, .85198)
\end{aligned}$$

and this array of iterates gives us about the same approximation to the fixed point.

11.3 Fixed Points of Arbitrary Continuous Functions

In this section we are concerned with fixed points of arbitrary functions that are continuous on a given subset S of $\mathbf{R}^n$. We no longer require the functions to be contractions or to have any other special property. The most important classical theorem of this type is an extremely deep existence theorem that was discovered by Luitzen Egbertus Jan Brouwer in 1911 and is known as the **Brouwer fixed point theorem**.

11.3.1 The Brouwer Fixed Point Theorem

The Brouwer fixed point theorem can be stated as follows:

Suppose that S is a closed ball in $\mathbf{R}^n$ and that f is a continuous function from S to S then the function f must have at least one fixed point in S.

Note that if c is a point in $\mathbf{R}^n$ and r is a positive number then the closed ball with center c and radius r is defined to be the set

$$\{x \in \mathbf{R}^n \mid \|x - c\| \leq r\}.$$

In particular, if a and b are any two real numbers and $a \leq b$ then the closed interval $[a, b]$ is a closed ball in $\mathbf{R}^1$. In fact, if c is the midpoint of this interval and $r = \frac{b-a}{2}$ then

$$[a, b] = \{x \in \mathbf{R} \mid |x - c| \leq r\}.$$

We can therefore state a one dimensional special version of the Brouwer fixed point theorem as follows:

Suppose that a and b are real numbers and that $a \leq b$ and that f is a continuous function from $[a, b]$ to $[a, b]$. then the function f must have at least one fixed point in $[a, b]$.

This one dimensional form of the theorem is considerably easier to prove than the general result in $\mathbf{R}^n$.

The n-dimensional form of the Brouwer fixed point theorem can actually be improved a little. Instead of requiring the domain S to be a closed ball, it is sufficient to assume that S is any closed bounded convex subset of $\mathbf{R}^n$. A set S is called *closed* if it contains all of its boundary points and is said to be *convex* if given any two points x and y in S, the line segment that joins x to y is included in the set S.

So, for example, discs, (solid) triangles, rectangles and parallelograms are all convex subsets of $\mathbf{R}^2$. Intervals are convex subsets of $\mathbf{R}$. An example of a subset of $\mathbf{R}^2$ that is not convex is the set $\{(x, y) \in \mathbf{R}^2 \mid 1 \leq x^2 + y^2 \leq 4\}$.

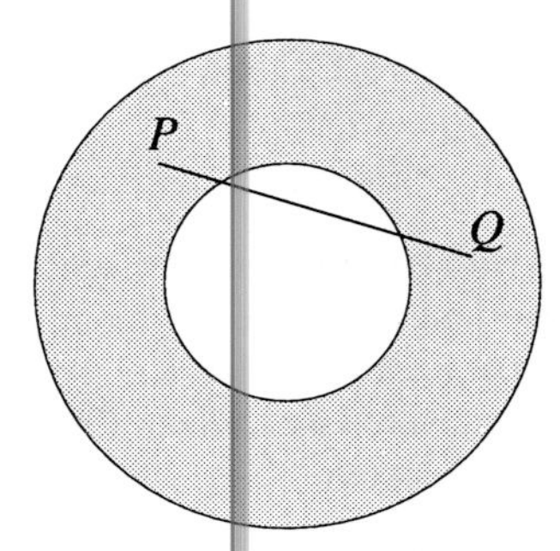

We can now restate the Brouwer fixed point theorem as follows;

Suppose that S is a closed bounded convex subset of $\mathbf{R}^n$ and that f is a continuous function from S to S then the function f must have at least one fixed point in S.

11.4 Exploring the Brouwer Theorem with *Scientific Notebook*

In this section we shall suggest a step by step procedure for obtaining successive approximations to a fixed point of a given continuous function f that is defined on a set S. While the method described here is not always the quickest or cleanest method of finding a fixed point of the given function, it does provide some insight into one of the ways in which the Brouwer fixed point theorem can be proved.

We shall focus on the method of proof of Brouwer's theorem that makes use of an important result, known as **Sperner's lemma** that was published in 1928 by Emanuel Sperner as part of Sperner's proof of another theorem of Brouwer known as the **dimensional invariance theorem**. It turned out that Sperner's lemma was also a valuable tool in the proof of Brouwer's fixed point theorem and it actually suggests a step by step algorithm for approximating the fixed point. Although the algorithm can be applied in the general n-dimensional case, we shall restrict our attention to the cases $n = 1$ and $n = 2$.[25]

11.4.1 The One Dimensional Case

11.4.1.1 Description of the Method

We shall describe a method for obtaining successive approximations to a fixed point of a continuous function $\phi : [a, b] \to [a, b]$. At each step in the process we shall be considering a subinterval $[u, v]$ of the interval $[a, b]$

[25] For more general discussion of this interesting topic, see the book *Invitation to Mathematics* by Konrad Jacobs, Princeton University Press, 1992.

$a \qquad u \quad v \qquad b$

in which the required fixed point lies and we shall replace this interval $[u,v]$ either by its lower half $\left[u,\frac{u+v}{2}\right]$ or by its upper half $\left[\frac{u+v}{2},v\right]$ which we shall then rename as $[u,v]$ again. At the beginning of the process we shall take $u=a$ and $v=b$. Since each step in the process replaces the interval $[u,v]$ by a new one with half the length, after n steps, the length of the interval $[u,v]$ is$(b-a)/2^n$ and so both u and v are approximations to the required fixed point with an error that does not exceed $(b-a)/2^n$.

In order to describe the process in detail we require a definition: A number x in the interval $[a,b]$ is said to be **cool** to the lower endpoint a of $[a,b]$ if $\phi(x)\geq x$ and is said to be cool to the upper endpoint b of $[a,b]$ if $\phi(x)\leq x$. Note that a number x is a fixed point of the function ϕ if and only if x is cool both to a and to b. We note that a is cool to a and that b is cool to b.

We now state an added requirement of the interval $[u,v]$ at each step of the process. At least one of the numbers u and v must be cool to a and at least one of the numbers u and v must be cool to b. We can express this requirement by saying that the interval $[u,v]$ must be **overall cool**. The one dimensional case of Sperner's lemma guarantees that if the interval $[u,v]$ is overall cool then either its lower half or its upper half must be overall cool.

11.4.1.2 Details of the Method

1. We begin by supplying the definitions of the numbers a and b to *Scientific Notebook*. For example, if $a=2$ then we point at the equation $a=2$ and click on Define and New Definition.
2. Supply the definition of the function $\phi:[a,b]\rightarrow[a,b]$ to *Scientific Notebook* by pointing at the formula for $\phi(x)$ and clicking on Define and New Definition.
3. In order to perform the selection process that selects either the upper or lower half of the interval $[u,v]$ at each step, supply the definitions

$$s(x)=\frac{x}{|x|}$$

and

$$f(u,v)=\frac{u+v}{2}+\left(\frac{u-v}{4}\right)\left|s\left(\phi\left(\frac{u+v}{2}\right)-\frac{u+v}{2}\right)-s(\phi(u)-u)\right|$$

to *Scientific Notebook*. Note that $s(x)=1$ whenever $x>0$ and $s(x)=-1$ whenever $x<0$. You can now see that if the interval $\left[u,\frac{u+v}{2}\right]$ is overall cool then

$$f(u,v)=\frac{u+v}{2}+\left(\frac{u-v}{4}\right)2=u$$

and if the interval $\left[u,\frac{u+v}{2}\right]$ is not overall cool then

$$f(u,v)=\frac{u+v}{2}+\left(\frac{u-v}{4}\right)0=\frac{u+v}{2}.$$

Now supply the definitions

$$g(u,v) = f(u,v) + \frac{v-u}{2}$$

and

$$h(u,v) = (f(u,v), g(u,v))$$

to *Scientific Notebook*. It is easy to see that the interval running from $f(u,v)$ to $g(u,v)$ is $\left[u, \frac{u+v}{2}\right]$ when $\left[u, \frac{u+v}{2}\right]$ is overall cool and is $\left[\frac{u+v}{2}, v\right]$ when $\left[u, \frac{u+v}{2}\right]$ is not overall cool.

4. Begin the step by step process by supplying the definitions $u = a$ and $v = b$. Point at the expression $h(u,v)$ and click on Evaluate Numerically. Then type the symbol h and drag the previously evaluated expression $h(u,v)$ to the right of the symbol h and click on Evaluate Numerically again. Again type h and drag the previously evaluated pair to the right of the symbol h and click on Evaluate Numerically. Repeat the process until your approximation is as close as you want it to be.

11.4.1.3 An Illustrative Example

In this example we shall take

$$\phi(x) = 1 + 2\ln x$$

for every $x \in [2,5]$. We saw in Subsection 11.1.2 that this function has a fixed point that lies between 3 and 4. After you have made your definitions as described above, if you click on Define and Show Definitions, you should see

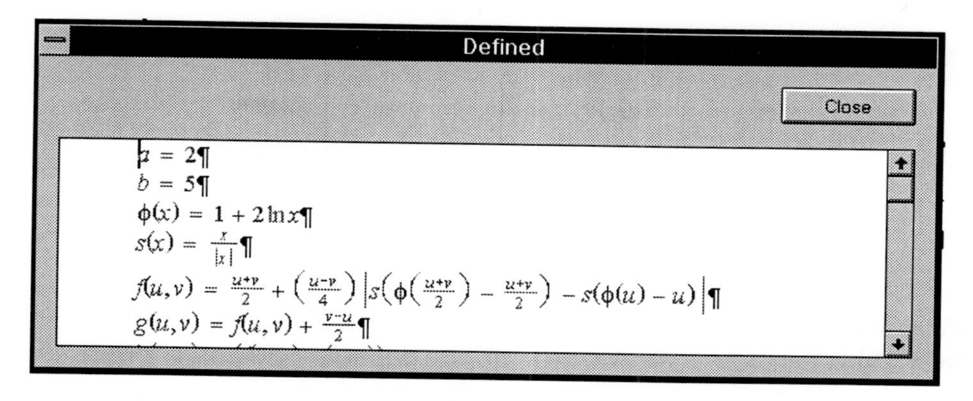

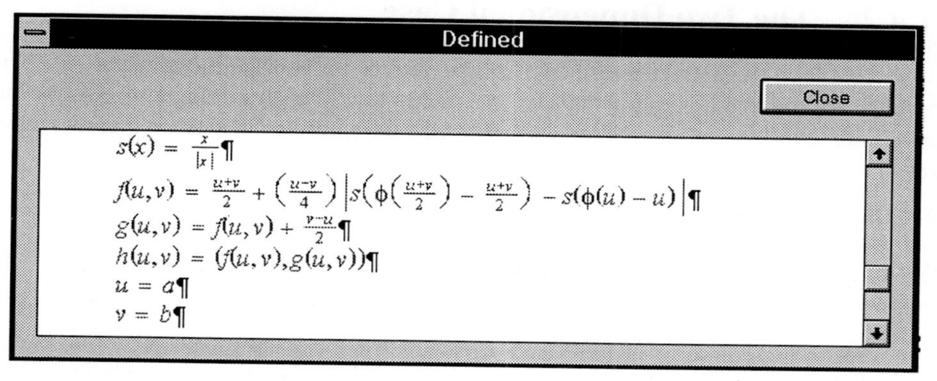

We had to show the list of definitions twice here in order to make them all visible. Now point at $h(u,v)$ and evaluate numerically. You will see

$$h(u,v) = (3.5, 5.0)$$

Type h and drag the pair $(3.5, 5.0)$ to its right and evaluate numerically again. You will see

$$h(3.5, 5.0) = (3.5, 4.25).$$

Repeat the process again and again to obtain

$$\begin{aligned}
h(3.5, 4.25) &= (3.5, 3.875) \\
h(3.5, 3.875) &= (3.5, 3.6875) \\
h(3.5, 3.6875) &= (3.5, 3.5938) \\
h(3.5, 3.5938) &= (3.5, 3.5469) \\
h(3.5, 3.5469) &= (3.5, 3.5235) \\
h(3.5, 3.5235) &= (3.5118, 3.5235) \\
h(3.5118, 3.5235) &= (3.5118, 3.5177) \\
h(3.5118, 3.5177) &= (3.5118, 3.5148) \\
h(3.5118, 3.5148) &= (3.5118, 3.5133)
\end{aligned}$$

$$\begin{aligned}
h(3.5118, 3.5133) &= (3.5126, 3.5133) \\
h(3.5126, 3.5133) &= (3.5126, 3.513) \\
h(3.5126, 3.513) &= (3.5128, 3.513) \\
h(3.5128, 3.513) &= (3.5128, 3.5129) \\
h(3.5128, 3.5129) &= (3.5129, 3.5129)
\end{aligned}$$

We deduce that the number 3.5129 approximates a fixed point of the function ϕ to four decimal places.

11.4.2 The Two Dimensional Case

As we have said, the method described in this subsection is not being presented as an efficient approach for calculating approximations to a fixed point of a given function. It is meant to provide some insight into one of the ways in which the Brouwer fixed point theorem may be proved.

11.4.2.1 Description of the Method

We shall describe a method for obtaining successive approximations to a fixed point of a continuous function $\phi : S \to S$ where S is the solid triangle with vertices

$$\begin{aligned}
A &= (0,0) \\
B &= (0,1) \\
C &= (1,0).
\end{aligned}$$

At the beginning of the process we shall take

$$\begin{aligned}
P_1 &= A \\
P_2 &= B \\
P_3 &= C.
\end{aligned}$$

The first step in the process is to partition the triangle $\triangle P_1P_2P_3$ into the four triangles $\triangle P_1M_2M_3$, $\triangle P_2M_3M_1$, $\triangle P_3M_1M_2$ and $\triangle M_1M_2M_3$ where M_1, M_2 and M_3 are, respectively, the midpoints of the sides P_2P_3, P_3P_1 and P_1P_2 of $\triangle P_1P_2P_3$.

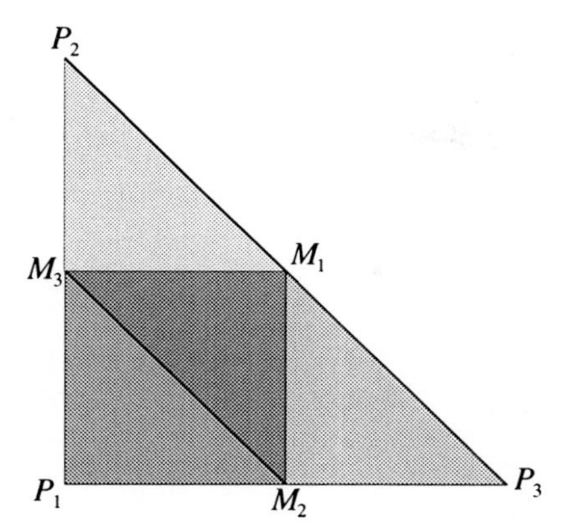

We shall then select one of these four smaller triangles and rename it $\triangle P_1 P_2 P_3$. Then we partition the new $\triangle P_1 P_2 P_3$ into four smaller triangles. The following figure illustrates one of the four possible ways in which step two may be performed.

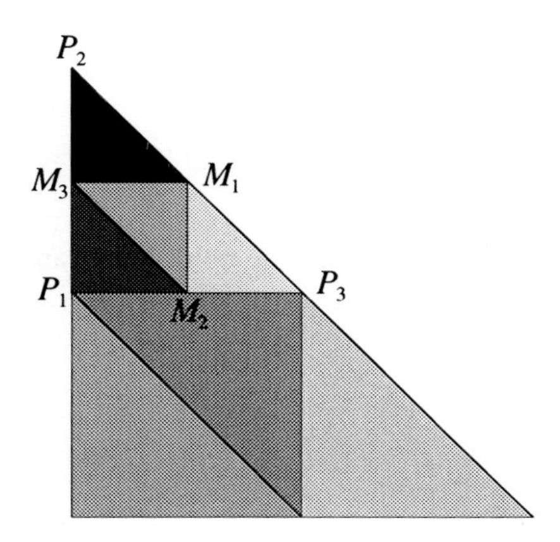

Repeating this procedure again and again, we obtain a contracting sequence of triangles that shrinks to a single point. If these triangles are chosen correctly at each step in the process then the point to which the sequence converges will be a fixed point of the given function. The following figure illustrates how the figure may appear after four steps of the procedure have been performed.

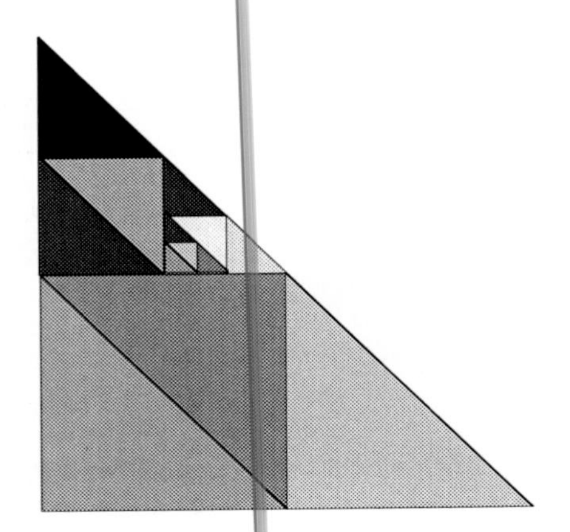

In order to describe the process in detail we require a definition: A point P in $\triangle ABC$ is said to be **cool** to the vertex A of $\triangle ABC$ if the distance from the point $\phi(P)$ to the line segment BC does not exceed the distance from P to BC. Similarly, a point P is said to be cool to the vertex B of $\triangle ABC$ if the distance from the point $\phi(P)$ to the line segment CA does not exceed the distance from P to CA and a point P is cool to C when the distance from the point $\phi(P)$ to the line segment AB does not exceed the distance from P to AB. Note that a point P is a fixed point of the function ϕ if and only if P is cool both to all three of the vertices A, B and C. Note also that each of the vertices A, B and C of $\triangle ABC$ is cool to itself.

We now state an added requirement of the triangle $\triangle P_1P_2P_3$ at each step of the process. We require that at least one of the points P_1, P_2 and P_3 must be cool to A, at least one of the points P_1, P_2 and P_3 must be cool to B and at least one of the points P_1, P_2 and P_3 must be cool to C. We can express this requirement by saying that the triangle $\triangle P_1P_2P_3$ must be *overall cool*. A special case of Sperner's lemma guarantees that if the triangle $\triangle P_1P_2P_3$ is overall cool then at least one of the four triangles into which we are subdividing it is also overall cool.

11.4.2.2 Details of the Method

1. We begin by supplying the definitions of the points A, B and C to *Scientific Notebook*. In other words, point at each of the equations

$$A = (0,0)$$

$$B = (0,1)$$

$$C = (1,0)$$

   and click on Define and New Definition.
2. We now supply the definition of the function ϕ to *Scientific Notebook* by pointing at the formula for $\phi(x,y)$ and clicking on Define and New Definition.
3. At each step in the procedure we are going to subdivide $\triangle P_1P_2P_3$ into the four

triangles $\Delta P_1M_2M_3$, $\Delta P_2M_3M_1$, $\Delta P_3M_1M_2$ and $\Delta M_1M_2M_3$ where M_1, M_2 and M_3 are, respectively, the midpoints of the sides P_2P_3, P_3P_1 and P_1P_2 of $\Delta P_1P_2P_3$. Accordingly, we supply the definitions

$$M_1 = \frac{1}{2}(P_2 + P_3)$$

$$M_2 = \frac{1}{2}(P_3 + P_1)$$

$$M_3 = \frac{1}{2}(P_1 + P_2)$$

to *Scientific Notebook*.

4. In order to measure whether or not a given point (x, y) is cool to the vertices A, B and C we need to supply *Scientific Notebook* with the distances from (x, y) to the line segments BC, CA and AB. We shall write these distances as $d_a(x, y)$, $d_b(x, y)$ and $d_c(x, y)$. We use lower case letters a, b and c for these suffixes because the upper case symbols A, B and C are already in use. Accordingly, we point at each of the equations

$$d_c(x, y) = \left\| \left(1 + \frac{(A - (x, y)) \cdot (B - A)}{\|B - A\|^2}\right) A - \frac{(A - (x, y)) \cdot (B - A)}{\|B - A\|^2} B - (x, y) \right\|$$

$$d_a(x, y) = \left\| \left(1 + \frac{(B - (x, y)) \cdot (C - B)}{\|C - B\|^2}\right) B - \frac{(B - (x, y)) \cdot (C - B)}{\|C - B\|^2} C - (x, y) \right\|$$

$$d_b(x, y) = \left\| \left(1 + \frac{(C - (x, y)) \cdot (A - C)}{\|A - C\|^2}\right) C - \frac{(C - (x, y)) \cdot (A - C)}{\|A - C\|^2} A - (x, y) \right\|$$

and click on Define and New Definition.

5. We now supply *Scientific Notebook* with the expressions that tell us whether a given point (x, y) is cool to A, B or C. We point at the equations

$$g_a(x, y) = d_a(x, y) - d_a\phi(x, y)$$

$$g_b(x, y) = d_b(x, y) - d_b\phi(x, y)$$

$$g_c(x, y) = d_c(x, y) - d_c\phi(x, y)$$

and click on Define and New Definition. The condition for a point (x, y) to be cool to A is that $g_a(x, y) \geq 0$. The point (x, y) is cool to B when $g_b(x, y) \geq 0$ and is cool to C when $g_c(x, y) \geq 0$.

6. We now describe the procedure that we employ at each step of the process. We fill the coordinates of the points P_1, P_2 and P_3 into the array

P_1	$x_1 =$	$y_1 =$
P_2	$x_2 =$	$y_2 =$
P_3	$x_3 =$	$y_3 =$
M_1	$u_1 =$	$v_1 =$
M_2	$u_2 =$	$v_2 =$
M_3	$u_3 =$	$v_3 =$

Thus, for example, the first of these arrays will be filled as

$P_1 = (0,0)$	$x_1 = 0$	$y_1 = 0$
$P_2 = (0,1)$	$x_2 = 0$	$y_2 = 1$
$P_3 = (1,0)$	$x_3 = 1$	$y_3 = 0$
M_1	$u_1 =$	$v_1 =$
M_2	$u_2 =$	$v_2 =$
M_3	$u_3 =$	$v_3 =$

Then we point at each of the nine equations in the top three rows of this array and click on Define and New Definition and we then evaluate each of the points M_1, M_2 and M_3. When you evaluate a point such as M_1, you may see $\frac{1}{2}P_2 + \frac{1}{2}P_3$. If so, just highlight the latter expression, hold down your control button and click on Evaluate again. Now drag the coordinates of M_1 into the positions reserved for the values of u_1 and v_1 and fill in the values of u_2, v_2, u_3 and v_3 in the same way. The first of these arrays will therefore appear as

$P_1 = (0,0)$	$x_1 = 0$	$y_1 = 0$
$P_2 = (0,1)$	$x_2 = 0$	$y_2 = 1$
$P_3 = (1,0)$	$x_3 = 1$	$y_3 = 0$
$M_1 = \left(\frac{1}{2}, \frac{1}{2}\right)$	$u_1 = \frac{1}{2}$	$v_1 = \frac{1}{2}$
$M_2 = \left(\frac{1}{2}, 0\right)$	$u_2 = \frac{1}{2}$	$v_2 = 0$
$M_3 = \left(0, \frac{1}{2}\right)$	$u_3 = 0$	$v_3 = \frac{1}{2}$

Point at the equations that define u_1, v_1, u_2, v_2, u_3 and v_3 and click on Define and New

Definition. Then evaluate the eighteen values of the functions g_a, g_b and g_c in the array

	A	B	C
P_1	$g_a(x_1, y_1)$	$g_b(x_1, y_1)$	$g_c(x_1, y_1)$
P_2	$g_a(x_2, y_2)$	$g_b(x_2, y_2)$	$g_c(x_2, y_2)$
P_3	$g_a(x_3, y_3)$	$g_b(x_3, y_3)$	$g_c(x_3, y_3)$
M_1	$g_a(u_1, v_1)$	$g_b(u_1, v_1)$	$g_c(u_1, v_1)$
M_2	$g_a(u_2, v_2)$	$g_b(u_2, v_2)$	$g_c(u_2, v_2)$
M_3	$g_a(u_3, v_3)$	$g_b(u_3, v_3)$	$g_c(u_3, v_3)$

Look at the values you have obtained and choose one triangle among the four triangles $\triangle P_1M_2M_3$, $\triangle P_2M_3M_1$, $\triangle P_3M_1M_2$ and $\triangle M_1M_2M_3$ that is overall cool. Take the vertices of this triangle as the new values of the points P_1, P_2 and P_3 and repeat the procedure.

11.4.2.3 An Illustrative Example

In this example we shall find successive approximations to a fixed point of the function ϕ defined by

$$\phi(x, y) = \left(\left|\frac{1}{2} - x\right| |\cos y|, \left|\frac{1}{2} - y\right| |\cos x|\right)$$

at every point $(x, y) \in \triangle ABC$. It is easy to see that ϕ is a continuous function from $\triangle ABC$ into $\triangle ABC$.

Step 1

With the coordinates of the points P_1, P_2, P_3, M_1, M_2 and M_3 shown in the array

$P_1 = (0, 0)$	$x_1 = 0$	$y_1 = 0$
$P_2 = (0, 1)$	$x_2 = 0$	$y_2 = 1$
$P_3 = (1, 0)$	$x_3 = 1$	$y_3 = 0$
$M_1 = \left(\frac{1}{2}, \frac{1}{2}\right)$	$u_1 = \frac{1}{2}$	$v_1 = \frac{1}{2}$
$M_2 = \left(\frac{1}{2}, 0\right)$	$u_2 = \frac{1}{2}$	$v_2 = 0$
$M_3 = \left(0, \frac{1}{2}\right)$	$u_3 = 0$	$v_3 = \frac{1}{2}$

the array

	A	B	C
P_1	$g_a(x_1, y_1)$	$g_b(x_1, y_1)$	$g_c(x_1, y_1)$
P_2	$g_a(x_2, y_2)$	$g_b(x_2, y_2)$	$g_c(x_2, y_2)$
P_3	$g_a(x_3, y_3)$	$g_b(x_3, y_3)$	$g_c(x_3, y_3)$
M_1	$g_a(u_1, v_1)$	$g_b(u_1, v_1)$	$g_c(u_1, v_1)$
M_2	$g_a(u_2, v_2)$	$g_b(u_2, v_2)$	$g_c(u_2, v_2)$
M_3	$g_a(u_3, v_3)$	$g_b(u_3, v_3)$	$g_c(u_3, v_3)$

comes out as[26]

	A	B	C
P_1	$g_a(x_1, y_1) = .70711$	$g_b(x_1, y_1) = -.5$	$g_c(x_1, y_1) = -.5$
P_2	$g_a(x_2, y_2) = -.16253$	$g_b(x_2, y_2) = .5$	$g_c(x_2, y_2) = -.27015$
P_3	$g_a(x_3, y_3) = -.16253$	$g_b(x_3, y_3) = -.27015$	$g_c(x_3, y_3) = .5$
M_1	$g_a(u_1, v_1) = -.70711$	$g_b(u_1, v_1) = .5$	$g_c(u_1, v_1) = .5$
M_2	$g_a(u_2, v_2) = -4.3281 \times 10^{-2}$	$g_b(u_2, v_2) = -.43879$	$g_c(u_2, v_2) = .5$
M_3	$g_a(u_3, v_3) = -4.3281 \times 10^{-2}$	$g_b(u_3, v_3) = .5$	$g_c(u_3, v_3) = -.43879$

and we can see from this array that among the four triangles $\triangle P_1M_2M_3$, $\triangle P_2M_3M_1$, $\triangle P_3M_1M_2$ and $\triangle M_1M_2M_3$, only the triangle $\triangle P_1M_2M_3$ is overall cool. Accordingly, we keep the point P_1 as $(0, 0)$ but replace the coordinates of P_2 and P_3 by those of M_2 and M_3 respectively.

Step 2

With the coordinates of the points P_1, P_2, P_3, M_1, M_2 and M_3 shown in the array

$P_1 = (0, 0)$	$x_1 = 0$	$y_1 = 0$
$P_2 = (\frac{1}{2}, 0)$	$x_2 = \frac{1}{2}$	$y_2 = 0$
$P_3 = (0, \frac{1}{2})$	$x_3 = 0$	$y_3 = \frac{1}{2}$
$M_1 = (\frac{1}{4}, \frac{1}{4})$	$u_1 = \frac{1}{4}$	$v_1 = \frac{1}{4}$
$M_2 = (0, \frac{1}{4})$	$u_2 = 0$	$v_2 = \frac{1}{4}$
$M_3 = (\frac{1}{4}, 0)$	$u_3 = \frac{1}{4}$	$v_3 = 0$

the array

26 The following expression will be fully visible only in the on-screen version of this text.

	A	B	C
P_1	$g_a(x_1, y_1)$	$g_b(x_1, y_1)$	$g_c(x_1, y_1)$
P_2	$g_a(x_2, y_2)$	$g_b(x_2, y_2)$	$g_c(x_2, y_2)$
P_3	$g_a(x_3, y_3)$	$g_b(x_3, y_3)$	$g_c(x_3, y_3)$
M_1	$g_a(u_1, v_1)$	$g_b(u_1, v_1)$	$g_c(u_1, v_1)$
M_2	$g_a(u_2, v_2)$	$g_b(u_2, v_2)$	$g_c(u_2, v_2)$
M_3	$g_a(u_3, v_3)$	$g_b(u_3, v_3)$	$g_c(u_3, v_3)$

comes out as[27]

	A	B	C
P_1	$g_a(x_1, y_1) = .70711$	$g_b(x_1, y_1) = -.5$	$g_c(x_1, y_1) = -.5$
P_2	$g_a(x_2, y_2) = -4.3281 \times 10^{-2}$	$g_b(x_2, y_2) = -.43879$	$g_c(x_2, y_2) = .5$
P_3	$g_a(x_3, y_3) = -4.3281 \times 10^{-2}$	$g_b(x_3, y_3) = .5$	$g_c(x_3, y_3) = -.43879$
M_1	$g_a(u_1, v_1) = -1.0991 \times 10^{-2}$	$g_b(u_1, v_1) = 7.7719 \times 10^{-3}$	$g_c(u_1, v_1) = 7.7719 \times 10^{-3}$
M_2	$g_a(u_2, v_2) = .34256$	$g_b(u_2, v_2) = 0$	$g_c(u_2, v_2) = -.48446$
M_3	$g_a(u_3, v_3) = .34256$	$g_b(u_3, v_3) = -.48446$	$g_c(u_3, v_3) = 0$

and we can see from this array that all four of the triangles $\triangle P_1M_2M_3$, $\triangle P_2M_3M_1$, $\triangle P_3M_1M_2$ and $\triangle M_1M_2M_3$ are overall cool. We can therefore choose any one of the four triangles as our new triangle $\triangle P_1P_2P_3$. Rather arbitrarily, we choose $\triangle P_3M_1M_2$ as our new triangle.

Step 3

With the coordinates of the points P_1, P_2, P_3, M_1, M_2 and M_3 shown in the array

$P_1 = \left(\frac{1}{4}, \frac{1}{4}\right)$	$x_1 = \frac{1}{4}$	$y_1 = \frac{1}{4}$
$P_2 = \left(0, \frac{1}{4}\right)$	$x_2 = 0$	$y_2 = \frac{1}{4}$
$P_3 = \left(0, \frac{1}{2}\right)$	$x_3 = 0$	$y_3 = \frac{1}{2}$
$M_1 = \left(0, \frac{3}{8}\right)$	$u_1 = 0$	$v_1 = \frac{3}{8}$
$M_2 = \left(\frac{1}{8}, \frac{3}{8}\right)$	$u_2 = \frac{1}{8}$	$v_2 = \frac{3}{8}$
$M_3 = \left(\frac{1}{8}, \frac{1}{4}\right)$	$u_3 = \frac{1}{8}$	$v_3 = \frac{1}{4}$

the array

[27] The following expression will be fully visible only in the on-screen version of this text.

	A	B	C
P_1	$g_a(x_1, y_1)$	$g_b(x_1, y_1)$	$g_c(x_1, y_1)$
P_2	$g_a(x_2, y_2)$	$g_b(x_2, y_2)$	$g_c(x_2, y_2)$
P_3	$g_a(x_3, y_3)$	$g_b(x_3, y_3)$	$g_c(x_3, y_3)$
M_1	$g_a(u_1, v_1)$	$g_b(u_1, v_1)$	$g_c(u_1, v_1)$
M_2	$g_a(u_2, v_2)$	$g_b(u_2, v_2)$	$g_c(u_2, v_2)$
M_3	$g_a(u_3, v_3)$	$g_b(u_3, v_3)$	$g_c(u_3, v_3)$

comes out as[28]

	A	B	C
P_1	$g_a(x_1, y_1) = -1.0991 \times 10^{-2}$	$g_b(x_1, y_1) = 7.7719 \times 10^{-3}$	$g_c(x_1, y_1) = 7.7719 \times 10^{-}$
P_2	$g_a(x_2, y_2) = .34256$	$g_b(x_2, y_2) = 0$	$g_c(x_2, y_2) = -.48446$
P_3	$g_a(x_3, y_3) = -4.3281 \times 10^{-2}$	$g_b(x_3, y_3) = .5$	$g_c(x_3, y_3) = -.43879$
M_1	$g_a(u_1, v_1) = .15221$	$g_b(u_1, v_1) = .25$	$g_c(u_1, v_1) = -.46525$
M_2	$g_a(u_2, v_2) = -1.9117 \times 10^{-2}$	$g_b(u_2, v_2) = .25098$	$g_c(u_2, v_2) = -.22394$
M_3	$g_a(u_3, v_3) = .16715$	$g_b(u_3, v_3) = 1.9506 \times 10^{-3}$	$g_c(u_3, v_3) = -.23834$

We observe that $\triangle P_1 M_2 M_3$ is overall cool and we choose it for our next triangle.

Step 4

With the coordinates of the points P_1, P_2, P_3, M_1, M_2 and M_3 shown in the array

$P_1 = \left(\frac{1}{4}, \frac{1}{4}\right)$	$x_1 = \frac{1}{4}$	$y_1 = \frac{1}{4}$
$P_2 = \left(\frac{1}{8}, \frac{3}{8}\right)$	$x_2 = \frac{1}{8}$	$y_2 = \frac{3}{8}$
$P_3 = \left(\frac{1}{8}, \frac{1}{4}\right)$	$x_3 = \frac{1}{8}$	$y_3 = \frac{1}{4}$
$M_1 = \left(\frac{1}{8}, \frac{5}{16}\right)$	$u_1 = \frac{1}{8}$	$v_1 = \frac{5}{16}$
$M_2 = \left(\frac{3}{16}, \frac{1}{4}\right)$	$u_2 = \frac{3}{16}$	$v_2 = \frac{1}{4}$
$M_3 = \left(\frac{3}{16}, \frac{5}{16}\right)$	$u_3 = \frac{3}{16}$	$v_3 = \frac{5}{16}$

the array

[28] The following expression will be fully visible only in the on-screen version of this text.

	A	B	C
P_1	$g_a(x_1,y_1)$	$g_b(x_1,y_1)$	$g_c(x_1,y_1)$
P_2	$g_a(x_2,y_2)$	$g_b(x_2,y_2)$	$g_c(x_2,y_2)$
P_3	$g_a(x_3,y_3)$	$g_b(x_3,y_3)$	$g_c(x_3,y_3)$
M_1	$g_a(u_1,v_1)$	$g_b(u_1,v_1)$	$g_c(u_1,v_1)$
M_2	$g_a(u_2,v_2)$	$g_b(u_2,v_2)$	$g_c(u_2,v_2)$
M_3	$g_a(u_3,v_3)$	$g_b(u_3,v_3)$	$g_c(u_3,v_3)$

comes out as[29]

	A	B	C
P_1	$g_a(x_1,y_1) = -1.0991 \times 10^{-2}$	$g_b(x_1,y_1) = 7.7719 \times 10^{-3}$	$g_c(x_1,y_1) = 7.7719 \times 10^{-3}$
P_2	$g_a(x_2,y_2) = -1.9117 \times 10^{-2}$	$g_b(x_2,y_2) = .25098$	$g_c(x_2,y_2) = -.22394$
P_3	$g_a(x_3,y_3) = .16715$	$g_b(x_3,y_3) = 1.9506 \times 10^{-3}$	$g_c(x_3,y_3) = -.23834$
M_1	$g_a(u_1,v_1) = 7.4511 \times 10^{-2}$	$g_b(u_1,v_1) = .12646$	$g_c(u_1,v_1) = -.23184$
M_2	$g_a(u_2,v_2) = 7.8421 \times 10^{-2}$	$g_b(u_2,v_2) = 4.3817 \times 10^{-3}$	$g_c(u_2,v_2) = -.11529$
M_3	$g_a(u_3,v_3) = -1.3026 \times 10^{-2}$	$g_b(u_3,v_3) = .12829$	$g_c(u_3,v_3) = -.10986$

We observe that $\triangle P_1 M_2 M_3$ is overall cool and we choose it for our next triangle.
Step 5

$P_1 = \left(\frac{1}{4}, \frac{1}{4}\right)$	$x_1 = \frac{1}{4}$	$y_1 = \frac{1}{4}$
$P_2 = \left(\frac{3}{16}, \frac{1}{4}\right)$	$x_2 = \frac{3}{16}$	$y_2 = \frac{1}{4}$
$P_3 = \left(\frac{3}{16}, \frac{5}{16}\right)$	$x_3 = \frac{3}{16}$	$y_3 = \frac{5}{16}$
$M_1 = \left(\frac{3}{16}, \frac{9}{32}\right)$	$u_1 = \frac{3}{16}$	$v_1 = \frac{9}{32}$
$M_2 = \left(\frac{7}{32}, \frac{9}{32}\right)$	$u_2 = \frac{7}{32}$	$v_2 = \frac{9}{32}$
$M_3 = \left(\frac{7}{32}, \frac{1}{4}\right)$	$u_3 = \frac{7}{32}$	$v_3 = \frac{1}{4}$

[29] The following expression will be fully visible only in the on-screen version of this text.

	A	B	C
P_1	$g_a(x_1, y_1) = -1.0991 \times 10^{-2}$	$g_b(x_1, y_1) = 7.7719 \times 10^{-3}$	$g_c(x_1, y_1) = 7.7719 \times 10^{-}$
P_2	$g_a(x_2, y_2) = 7.8421 \times 10^{-2}$	$g_b(x_2, y_2) = 4.3817 \times 10^{-3}$	$g_c(x_2, y_2) = -.11529$
P_3	$g_a(x_3, y_3) = -1.3026 \times 10^{-2}$	$g_b(x_3, y_3) = .12829$	$g_c(x_3, y_3) = -.10986$
M_1	$g_a(u_1, v_1) = 3.2801 \times 10^{-2}$	$g_b(u_1, v_1) = 6.6334 \times 10^{-2}$	$g_c(u_1, v_1) = -.11272$
M_2	$g_a(u_2, v_2) = -.0115$	$g_b(u_2, v_2) = 6.7713 \times 10^{-2}$	$g_c(u_2, v_2) = -5.1449 \times 10$
M_3	$g_a(u_3, v_3) = 3.3799 \times 10^{-2}$	$g_b(u_3, v_3) = 5.9576 \times 10^{-3}$	$g_c(u_3, v_3) = -5.3757 \times 10$

We observe that $\triangle P_1 M_2 M_3$ is overall cool and we choose it for our next triangle.

Step 6

$P_1 = \left(\frac{1}{4}, \frac{1}{4}\right)$	$x_1 = \frac{1}{4}$	$y_1 = \frac{1}{4}$
$P_2 = \left(\frac{7}{32}, \frac{9}{32}\right)$	$x_2 = \frac{7}{32}$	$y_2 = \frac{9}{32}$
$P_3 = \left(\frac{7}{32}, \frac{1}{4}\right)$	$x_3 = \frac{7}{32}$	$y_3 = \frac{1}{4}$
$M_1 = \left(\frac{7}{32}, \frac{17}{64}\right)$	$u_1 = \frac{7}{32}$	$v_1 = \frac{17}{64}$
$M_2 = \left(\frac{15}{64}, \frac{1}{4}\right)$	$u_2 = \frac{15}{64}$	$v_2 = \frac{1}{4}$
$M_3 = \left(\frac{15}{64}, \frac{17}{64}\right)$	$u_3 = \frac{15}{64}$	$v_3 = \frac{17}{64}$

	A	B	C
P_1	$g_a(x_1, y_1) = -1.0991 \times 10^{-2}$	$g_b(x_1, y_1) = 7.7719 \times 10^{-3}$	$g_c(x_1, y_1) = 7.7719 \times 10^{-}$
P_2	$g_a(x_2, y_2) = -.0115$	$g_b(x_2, y_2) = 6.7713 \times 10^{-2}$	$g_c(x_2, y_2) = -5.1449 \times 10$
P_3	$g_a(x_3, y_3) = 3.3799 \times 10^{-2}$	$g_b(x_3, y_3) = 5.9576 \times 10^{-3}$	$g_c(x_3, y_3) = -5.3757 \times 10$
M_1	$g_a(u_1, v_1) = 1.1173 \times 10^{-2}$	$g_b(u_1, v_1) = 3.6835 \times 10^{-2}$	$g_c(u_1, v_1) = -5.2636 \times 10$
M_2	$g_a(u_2, v_2) = 1.1425 \times 10^{-2}$	$g_b(u_2, v_2) = 6.8351 \times 10^{-3}$	$g_c(u_2, v_2) = -2.2992 \times 10$
M_3	$g_a(u_3, v_3) = -1.1118 \times 10^{-2}$	$g_b(u_3, v_3) = 3.7658 \times 10^{-2}$	$g_c(u_3, v_3) = -2.1934 \times 10$

We observe that $\triangle P_1 M_2 M_3$ is overall cool and we choose it for our next triangle.

Step 7

$P_1 = \left(\frac{1}{4}, \frac{1}{4}\right)$	$x_1 = \frac{1}{4}$	$y_1 = \frac{1}{4}$
$P_2 = \left(\frac{15}{64}, \frac{1}{4}\right)$	$x_2 = \frac{15}{64}$	$y_2 = \frac{1}{4}$
$P_3 = \left(\frac{15}{64}, \frac{17}{64}\right)$	$x_3 = \frac{15}{64}$	$y_3 = \frac{17}{64}$
$M_1 = \left(\frac{15}{64}, \frac{33}{128}\right)$	$u_1 = \frac{15}{64}$	$v_1 = \frac{33}{128}$
$M_2 = \left(\frac{31}{128}, \frac{33}{128}\right)$	$u_2 = \frac{31}{128}$	$v_2 = \frac{33}{128}$
$M_3 = \left(\frac{31}{128}, \frac{1}{4}\right)$	$u_3 = \frac{31}{128}$	$v_3 = \frac{1}{4}$

	A	B	C
P_1	$g_a(x_1, y_1) = -1.0991 \times 10^{-2}$	$g_b(x_1, y_1) = 7.7719 \times 10^{-3}$	$g_c(x_1, y_1) = 7.7719 \times 10^{-3}$
P_2	$g_a(x_2, y_2) = 1.1425 \times 10^{-2}$	$g_b(x_2, y_2) = 6.8351 \times 10^{-3}$	$g_c(x_2, y_2) = -2.2992 \times 10^{-2}$
P_3	$g_a(x_3, y_3) = -1.1118 \times 10^{-2}$	$g_b(x_3, y_3) = 3.7658 \times 10^{-2}$	$g_c(x_3, y_3) = -2.1934 \times 10^{-2}$
M_1	$g_a(u_1, v_1) = 1.5883 \times 10^{-4}$	$g_b(u_1, v_1) = 2.2246 \times 10^{-2}$	$g_c(u_1, v_1) = -2.2471 \times 10^{-2}$
M_2	$g_a(u_2, v_2) = -1.1023 \times 10^{-2}$	$g_b(u_2, v_2) = 2.2693 \times 10^{-2}$	$g_c(u_2, v_2) = -7.1043 \times 10^{-3}$
M_3	$g_a(u_3, v_3) = 2.2214 \times 10^{-4}$	$g_b(u_3, v_3) = 7.2961 \times 10^{-3}$	$g_c(u_3, v_3) = -7.6102 \times 10^{-3}$

We observe that $\triangle P_1 M_2 M_3$ is overall cool and we choose it for our next triangle.

Step 8

$P_1 = \left(\frac{1}{4}, \frac{1}{4}\right)$	$x_1 = \frac{1}{4}$	$y_1 = \frac{1}{4}$
$P_2 = \left(\frac{31}{128}, \frac{33}{128}\right)$	$x_2 = \frac{31}{128}$	$y_2 = \frac{33}{128}$
$P_3 = \left(\frac{31}{128}, \frac{1}{4}\right)$	$x_3 = \frac{31}{128}$	$y_3 = \frac{1}{4}$
$M_1 = \left(\frac{31}{128}, \frac{65}{256}\right)$	$u_1 = \frac{31}{128}$	$v_1 = \frac{65}{256}$
$M_2 = \left(\frac{63}{256}, \frac{1}{4}\right)$	$u_2 = \frac{63}{256}$	$v_2 = \frac{1}{4}$
$M_3 = \left(\frac{63}{256}, \frac{65}{256}\right)$	$u_3 = \frac{63}{256}$	$v_3 = \frac{65}{256}$

	A	B	C
P_1	$g_a(x_1, y_1) = -1.0991 \times 10^{-2}$	$g_b(x_1, y_1) = 7.7719 \times 10^{-3}$	$g_c(x_1, y_1) = 7.7719 \times 10^{-3}$
P_2	$g_a(x_2, y_2) = -1.1023 \times 10^{-2}$	$g_b(x_2, y_2) = 2.2693 \times 10^{-2}$	$g_c(x_2, y_2) = -7.1043 \times 10^{-}$
P_3	$g_a(x_3, y_3) = 2.2214 \times 10^{-4}$	$g_b(x_3, y_3) = 7.2961 \times 10^{-3}$	$g_c(x_3, y_3) = -7.6102 \times 10^{-}$
M_1	$g_a(u_1, v_1) = -5.399 \times 10^{-3}$	$g_b(u_1, v_1) = 1.4995 \times 10^{-2}$	$g_c(u_1, v_1) = -7.3592 \times 10^{-}$
M_2	$g_a(u_2, v_2) = -5.3832 \times 10^{-3}$	$g_b(u_2, v_2) = 7.5321 \times 10^{-3}$	$g_c(u_2, v_2) = 8.083 \times 10^{-5}$
M_3	$g_a(u_3, v_3) = -1.0999 \times 10^{-2}$	$g_b(u_3, v_3) = 1.5227 \times 10^{-2}$	$g_c(u_3, v_3) = 3.2809 \times 10^{-4}$

We observe that $\triangle M_1 M_2 P_3$ is overall cool and we choose it for our next triangle.

Step 9

$P_1 = \left(\frac{31}{128}, \frac{65}{256}\right)$	$x_1 = \frac{31}{128}$	$y_1 = \frac{65}{256}$
$P_2 = \left(\frac{63}{256}, \frac{1}{4}\right)$	$x_2 = \frac{63}{256}$	$y_2 = \frac{1}{4}$
$P_3 = \left(\frac{31}{128}, \frac{1}{4}\right)$	$x_3 = \frac{31}{128}$	$y_3 = \frac{1}{4}$
$M_1 = \left(\frac{125}{512}, \frac{1}{4}\right)$	$u_1 = \frac{125}{512}$	$v_1 = \frac{1}{4}$
$M_2 = \left(\frac{31}{128}, \frac{129}{512}\right)$	$u_2 = \frac{31}{128}$	$v_2 = \frac{129}{512}$
$M_3 = \left(\frac{125}{512}, \frac{129}{512}\right)$	$u_3 = \frac{125}{512}$	$v_3 = \frac{129}{512}$

	A	B	C
P_1	$g_a(x_1, y_1) = -5.399 \times 10^{-3}$	$g_b(x_1, y_1) = 1.4995 \times 10^{-2}$	$g_c(x_1, y_1) = -7.3592 \times 10$[illegible]
P_2	$g_a(x_2, y_2) = -5.3793 \times 10^{-3}$	$g_b(x_2, y_2) = 7.5321 \times 10^{-3}$	$g_c(x_2, y_2) = 8.083 \times 10^{-5}$
P_3	$g_a(x_3, y_3) = 2.2214 \times 10^{-4}$	$g_b(x_3, y_3) = 7.2961 \times 10^{-3}$	$g_c(x_3, y_3) = -7.6102 \times 10$[illegible]
M_1	$g_a(u_1, v_1) = -2.5802 \times 10^{-3}$	$g_b(u_1, v_1) = 7.4136 \times 10^{-3}$	$g_c(u_1, v_1) = -3.7647 \times 10$[illegible]
M_2	$g_a(u_2, v_2) = -2.5881 \times 10^{-3}$	$g_b(u_2, v_2) = 1.1145 \times 10^{-2}$	$g_c(u_2, v_2) = -7.4852 \times 10$[illegible]
M_3	$g_a(u_3, v_3) = -5.3891 \times 10^{-3}$	$g_b(u_3, v_3) = 1.1262 \times 10^{-2}$	$g_c(u_3, v_3) = -3.6406 \times 10$[illegible]

Notice how the values of the functions g_a, g_b and g_c at the vertices of the triangle become closer to zero as we proceed through the above working steps. Any of the points is an approximation to within a distance of $\frac{\sqrt{2}}{2^9}$ of a fixed point of the given function f.

Appendix A
Miscellaneous Notes

The notes and remarks in this chapter are targets of hypertext links at various points in the text.

1. We assume that P is a regular transition matrix and that λ is an eigenvalue of P. Choose an eigenvector x of P corresponding to λ. For each n we have $P^n x = \lambda^n x$. Now if we write

$$Q = \lim_{n\to\infty} P^n$$

then $P^n x \to Qx$ as $n \to \infty$. In other words, $\lambda^n x \to Qx$ as $n \to \infty$. Since the vector x is nonzero, the sequence of vectors $\lambda^n x$ can only converge if either $\lambda = 1$ or $|\lambda| < 1$.

2. It is not hard to see that the given sequence is increasing and bounded. Call its limit a. Letting $n \to \infty$ in the equation

$$a_{n+1} = \sqrt[3]{\frac{6a_n + 1}{8}}$$

and simplifying, we obtain

$$8a^3 - 6a - 1 = 0$$

We now look for numbers θ for which the number $\cos\theta$ is a solution of this cubic equation. Substituting $a = \cos\theta$ we obtain

$$8\cos^3\theta - 6\cos\theta = 1$$

which we can write as $\cos 3\theta = \frac{1}{2}$. It is now easy to see that the solution of this cubic equation is $a = \cos\frac{\pi}{9}$ or $a = -\cos\frac{2\pi}{9}$ or $a = -\cos\frac{4\pi}{9}$. Since the number a must be positive we conclude that $a = \cos\frac{\pi}{9}$.

3. Write $g(x) = f(x) - x$ for all $x > 0$. For each x we have $g'(x) = \frac{bc}{x} - 1$. Since there is only one number x for which $g'(x) = 0$, and since Rolle's theorem guarantees that $g'(x)$ must be zero at least once between any two numbers x at which $g(x) = 0$, we conclude that there are at most two numbers x at which $g(x) = 0$.

Index

Scientific Notebook (time-locked version)

Installation Instructions for Windows® 95 and Windows NT® 4.0

1. Start Windows and insert the program CD into the CD-ROM drive.

2. If the drive doesn't start automatically, click the Windows Start button, choose Run, and type D:/setup.exe, where D is the letter of your CD-ROM drive.

3. When the Scientific Notebook installation panel appears, follow the instructions on the screen.

4. Double-click the (Scientific Notebook LOGO) to begin working in Scientific Notebook.

System Requirements:

Windows® 95 or Windows NT® 4.0, and a CD-ROM drive. You must have your own Web browser to access Scientific Notebook's Resource Center on the World Wide Web. The program can be installed in a minimal (slower) form requiring 2 MB of hard disk space, a medium form, or a complete form (fastest) that requires from 35 MB to 80 MB of hard disk space, depending on your file system. Hard disk requirements are smallest when you use the new Windows 95 32-bit FAT or Windows NT's NTFS.

Ordering Information:

This full-featured, time-locked version of Scientific Notebook will expire 30 days after installation.
Purchase your own copy of Scientific Notebook on CD-ROM for just $59.95. Just place your order online at http://scinotebook.tcisoft.com; or call toll-free to: 1-800-874-2383, or write or FAX your order to:
(Please include the for-sale product's ISBN: 0-534-34864-5)

Brooks/Cole Publishing Company
511 Forest Lodge Road
Pacific Grove, CA 93950-5098
(408) 373-0728; FAX (408) 375-6414
e-mail: info@brookscole.com
Internet: http://www.brookscole.com

International Thomson Publishing Education Group

You'll find more information about Scientific Notebook online at
http://scinotebook.tcisoft.com

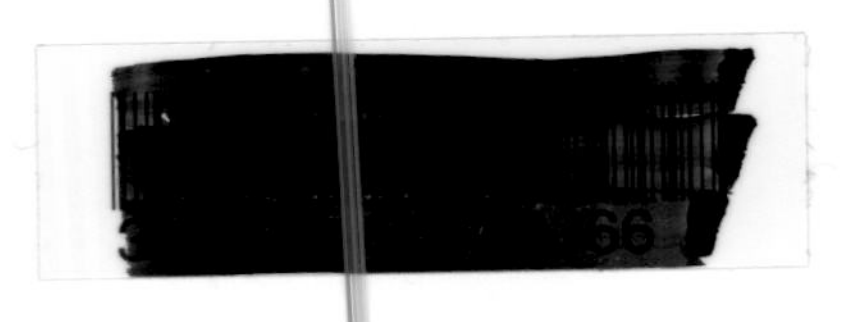